Finding Birds in Australia's Northern Territory

D Donato, P Wilkins, G Smith & L Alford

Publisher: Nick Alexander
Editor: Robin Taylor
Design & Layout: Tony Yap
Print Management: Roy Osborne
Cover Art: Di Skelly-Heron
Maps: Peter Lord
Photography:
Johnny Estbergs, Ian Morris, Raoul Slater,
Peter Wilkins, Environment Centre NT,
NT Tourist Commission

National Library of Australia
Cataloguing-in-Publication entry:
Finding birds in Australia's Northern Territory.
ISBN 0 643 06258 0
1. Birds - Northern Territory - Identification.
2. Bird watching - Northern Territory - Guidebooks.
I. Donato, David
598.072349429

Published by:
CSIRO PUBLISHING
150 Oxford Street
Collingwood, VIC 3066
Australia

Tel: (03) 9662 7666 Int: +(61) 3 9662 7666
Fax: (03) 9662 7555 Int: +(61) 3 9662 7555
Email: sales@publish.csiro.au
URL: http://www. publish.csiro.au

05 04 03 02 01 00 99 98 97
8 7 6 5 4 3 2 1

ACKNOWLEDGMENTS

Most of the information in this book comes from our own observations. Field trips into remote desert regions would not have been possible without the generous support of Ian Blevin of TREC Gondwana (see Appendix 1). In addition, his knowledge of birds and birdwatching in the arid Centre has greatly enhanced this book.

We also thank Phil Mitchell MLA and Deputy Chief Minister and Minister for Lands, Planning and Environment, Mike Reed for their support.

Andrew Ley provided editorial support. A friend in the field on many occasions, and a villain with his pen, his contributions are greatly appreciated.

Peter Lord from Ocean Art produced the maps. He had to put up with many alterations coming from four authors and always produced what we needed.

Di Skelly-Heron produced the artwork on the front cover. Her love of birds is reflected in her work. Easily recognisable and well known, her work is synonymous with the Northern Territory.

Jo Wilkins provided detailed research on the tourism and hospitality industry and also gave us her full support throughout the project.

We are grateful to the following for sharing their personal observations and birdwatching experiences with us: Richard Noske, Julian Reid, John Woinarski and Chris Materne. We also appreciated the help and advice of Ann Webb, and Bill and Jenny Green of Greenfingers Typesetting, Graphic Design and Publishing.

CONTENTS

AREA REFERENCE MAP

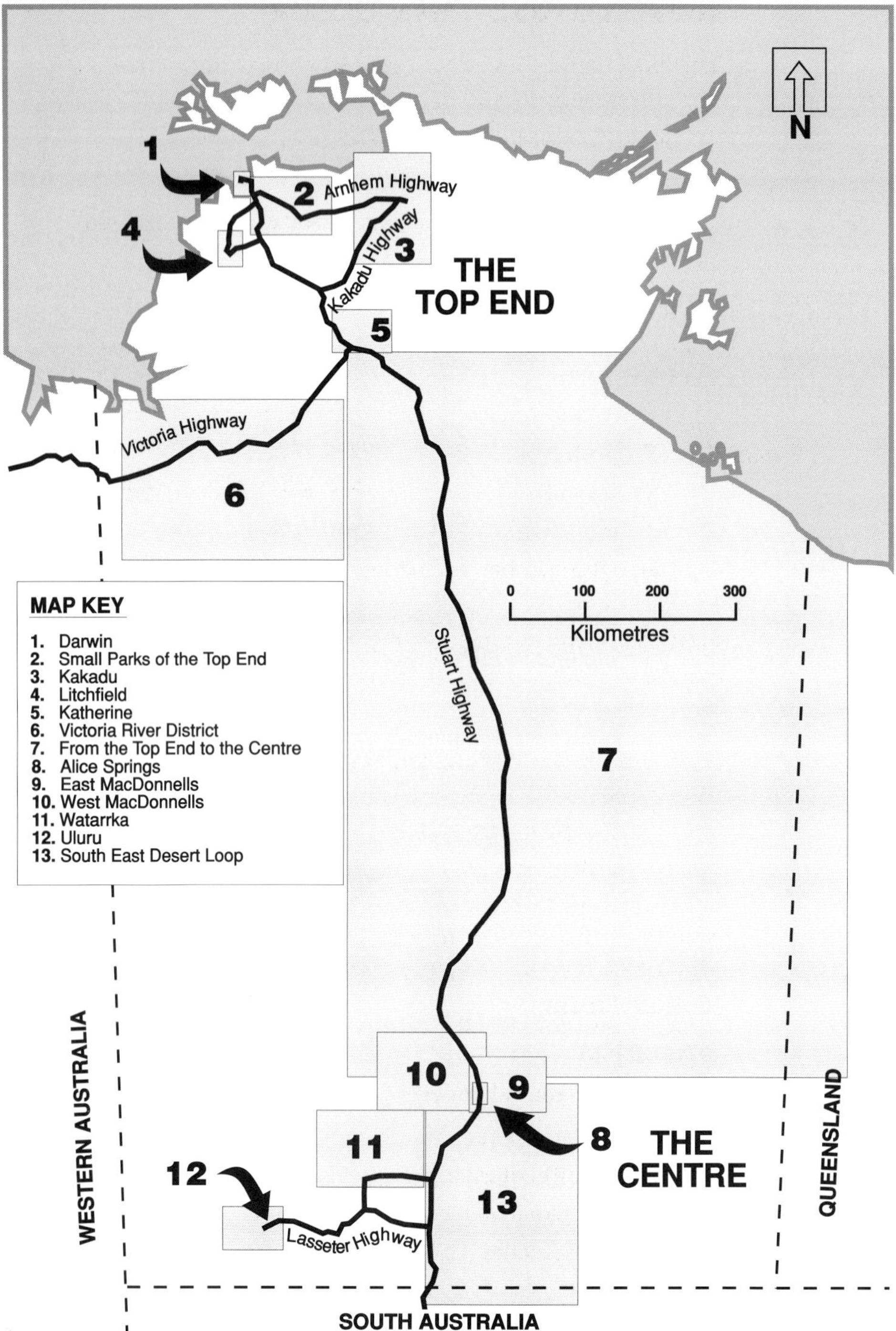

HOW TO USE THIS GUIDE

This book will lead you to the Northern Territory's best birdwatching areas and help you find the birds. It is designed to be used in conjunction with one of the Australian field identification guides (see Appendix 3).

The book has four main sections: **Habitats**, **Areas**, **Special Birds** and a **Checklist**.

The **Habitats** section describes broad classifications of habitats from a birdwatching point of view. They are not intended to be comprehensive botanical descriptions. Two sets of habitats are described — one for the Top End with its tropical climate and lush vegetation, and the other for the Centre with its arid climate, red sands, rugged ranges and hardy vegetation.

The **Areas** section has 13 chapters which describe the prime birdwatching sites in the Territory (see the map opposite). Each area chapter begins with a general description of the area followed by information on access, accommodation and camping.

Each area chapter has its own map which shows the location of the birdwatching sites as well as other useful features. The sites are listed in a logical order — either geographically (especially in the more confusing areas) or in order of the best birdwatching sites. The map is followed by a list of special birds found in the area and shows the recommended sites to look for each species.

The text does not attempt to discuss every bird likely to be found in every area, instead it points out interesting species found at each site.

The **Special Birds** section discusses birds considered to be special in the Northern Territory. It includes birds which are generally easier to find in the Territory than elsewhere in Australia. The discussion includes tips on finding the birds and, while not trying to be a field guide, it does include local identification hints. Special birds are also marked on the Checklist.

The **Checklist** contains all of the Territory's birds and the habitats and areas where they are likely to be found. It provides an easy way of accessing detailed information and avoids lengthy discussions about common birds in the text. "Twitchers" seeking to compile "hit lists" will find the Checklist extremely useful.

The bird names and sequence of species follow *The Taxonomy and Species of Birds of Australia and Its Territories* published by the Royal Australasian Ornithologists Union in 1994. Appendix 4 lists commonly used alternative names.

Appendix 1 lists organisations mentioned in the text. Appendix 2 lists the scientific names of plants mentioned throughout the book.

INTRODUCTION

Birds are a prominent feature of the Northern Territory environment. Almost half of Australia's bird species are found here in spectacular landscapes ranging from the desert to the tropics.

The Northern Territory is large (1.3 million square kilometres) and sparsely populated. Comprising one-sixth of the land mass it has only one percent of Australia's population. The large national parks, many square kilometres of open country and wide variety of habitats provide abundant birdwatching opportunities.

The Territory is generally divided into two regions, the northern Top End and the southern Centre. The Top End has a tropical climate with hot wet summers (the wet) and warm dry winters (the dry). The Centre has an arid climate with hot summers and mild winters. The winter months of June, July and August provide the most comfortable time to visit either region.

The Birds

Birdwatching in the Northern Territory is excellent all year. Many species that are greatly reduced in numbers elsewhere in Australia and South East Asia are still common, secure and easily observed here. Four species are currently considered to be endemic to the Territory. They are the Chestnut-quilled Rock Pigeon, the Banded Fruit-Dove, the Hooded Parrot and the White-throated Grasswren. A number of breeding migrants arrive from Asia during the summer months, and in winter the Territory receives an influx of birds from southern Australia. The 384 species listed in this book exclude only accidental visitors and pelagics.

The early mornings and late afternoons are best for successful birdwatching in the Northern Territory as the heat makes birds less active during the middle of the day.

Nocturnal birds are best seen during the first few hours after sunset.

Parks and Other Areas

The sites described are in parks or on other publicly accessible land. The parks are managed by the Parks and Wildlife Commission of the Northern Territory except for Uluru-Kata Tjuta and Kakadu, which are managed by Parks Australia North; some sites in towns are managed by local authorities.

The major roads in the Northern Territory are on public land. Feel free to pull up and wander about on the edge of the highways. Marked parking bays and picnic areas occur on the longer stretches and you can camp at some of these. Outside the park system most of the land is under pastoral lease or Aboriginal ownership and you need permission to enter. Contact the leaseholder or the relevant Land Council.

Most of the sites described are signposted and the parks have good interpretative material. Where things aren't so straightforward the text and maps contain additional information. Sites in the book are selected for their birdwatching qualities but many other areas and sites also provide excellent birdwatching opportunities — go exploring if you have time.

Travelling in the Northern Territory

Tourism is one of the Territory's major industries and plenty of modern facilities and infrastructure exist to accommodate visitors. The major centres of Darwin, Alice Springs, Katherine and Tennant Creek have tourism associations which will provide detailed information to help you plan your trip. Booking in advance is recommended. Darwin has an international and domestic airport and Alice Springs and Uluru have domestic airports with regular flights to other Australian centres.

Getting Around

To birdwatch successfully in the Territory, you will need your own vehicle, in most situations a two-wheel drive vehicle is all that is required. Where four-wheel drive is required, it is stated in the relevant chapter. Organised sightseeing tours visit some of the sites but usually they do not provide enough time for productive birdwatching. Specialist tours are available that cater for birdwatchers and general naturalists.

Accommodation and Camping

The major centres provide a full range of accommodation and most parks offer camping facilities. Daily fees range from a couple of dollars for basic facilities (water and pit toilets) to about ten dollars for campsites with flushing toilets and showers.

Comfort and Safety

In addition to binoculars, a field guide and this book, you will need a decent hat, 15+ sunscreen, sunglasses, insect repellent, good walking shoes, loose fitting clothes and lots of water (one litre per person per hour of walking is recommended). Knowledge of some first-aid and a first-aid kit and manual may also be useful. Care is needed to avoid dehydration and overheating and you should always notify rangers or police before undertaking long walks.

Mosquitoes can be a problem, particularly in the wetlands of the Top End. Unless signposts indicate otherwise, avoid swimming in waterways in the Top End due to Estuarine Crocodiles. Box Jellyfish (stingers) are common in Territory coastal waters from October to May and swimming in the ocean at this time is not recommended.

HABITATS

TOP END

Woodland

The most widespread habitat in the Top End, woodland has an overstorey dominated by eucalypts, primarily Woollybutt and Stringybark, and an understorey usually dominated by tall sorghum grasses. It also contains various other trees, shrubs, cycads, grasses and herbs. The habitat is reasonably uniform across its range but the structure and species composition may vary with changes in soil, topography, fire regime and frequency of flooding.

In most areas you will find a wide range of common bird species. However, subtle changes in the bird life will occur throughout the habitat and so it is worth visiting a number of woodland sites. For example, the Black-tailed Treecreeper, a fairly common woodland bird, is not uniformly distributed and is often absent from seemingly suitable habitat. Some families are well represented and a typical woodland site will have up to four species of pigeon, seven parrots, 12 honeyeaters, and a number of robins, whistlers and flycatchers. An increase in the number of tree species usually results in a wider variety and abundance of birds.

Flowering trees and shrubs attract nectar feeders. Woollybutt trees flower in the dry and their spectacular orange flowers attract many honeyeaters. Various eucalypts, grevilleas and other species can be found flowering throughout the year. Seed eaters are attracted to seeding grass, with finches and parrots being common in the more open grassed areas. Search the drier ridge areas for species such as Chestnut-backed Button-quail and Black-tailed Treecreeper. Also look for cuckoos — especially the Channel-billed Cuckoo if figs are present. Woodland habitats may have soaks or depressions housing some of the more versatile aquatic species, such as the Buff-banded Rail and White-faced Heron.

Monsoon Forest

The 'rainforest-like' monsoon forests are dark green in appearance with a diverse plant composition that is distinct from woodland. Located in fire protected areas around springs, waterbodies, waterways, escarpment valleys, coastal areas and rocky outcrops, they typically cover small, discrete areas of up to a couple of hectares. They generally have a closed canopy with a sparse understorey including ground ferns, palms and saplings. Epiphytes and climbing plants also grow there. Monsoon forests are sometimes referred to as vine-forests and when found in gorges at the edge of escarpments are called gallery forests. Their structure varies considerably depending on water availability and soil type, those around permanent water on the deep-soiled lowlands being taller and more diverse than those found in coastal areas and on rocky outcrops where they may be thicket-like in appearance.

Most bird species in the monsoon forests of the Northern Territory also occur in the rainforests of north Queensland, notable exceptions being the Banded Fruit-Dove and Rainbow Pitta. Typical species for this habitat include Orange-footed Scrubfowl, Green-backed Gerygone, White-gaped Honeyeater, Grey Whistler, Shining Flycatcher, Spangled Drongo, Varied Triller and Yellow Oriole.

Birds can be hard to find in monsoon forests as many stay in the canopy foliage and they are often shy and retiring. They are often more numerous and easy to see at the forest edge. Some species, however, such as the Rainbow Pitta, will only be seen inside the forest — listen for them scratching in the leaf litter. The Banded Fruit-Dove is endemic to the monsoon forests of the escarpment edge.

Escarpment

An escarpment occurs at the edge of a sandstone plateau. It is a rocky cliff-face that supports hardy woodland in the better soiled areas, and spinifex in the drier, more rocky parts. The plateau behind is covered by woodland vegetation, while waterfalls and gorges at the escarpment base support monsoon forests. The western escarpment of the Arnhem Land plateau extends from Kakadu National Park through to the Katherine region; escarpment habitat also occurs in the Victoria River District. Access to the escarpment can be difficult and one of the best sites to view the habitat is at Gunlom in Kakadu.

The escarpment contains many common woodland species and is also the sole habitat of a number of species. The Chestnut-quilled Rock-Pigeon (Kakadu) and White-quilled Rock-Pigeon (Victoria River District) can be found on the edge and top of the escarpment, as can the Sandstone Shrike-thrush and lavender-flanked form of the Variegated Fairy-wren. The White-throated Grasswren only occurs on top of the Arnhem Land escarpment in association with spinifex. Watch for Banded Fruit-Dove, Helmeted (Sandstone) Friarbird and White-lined Honeyeater flying around the ravines and escarpment edge. You may see cormorants and herons at rock pools.

Mangroves

A belt of mangroves extends along much of the Top End coast and along many tidal creeks and rivers. This habitat may extend for a considerable distance inland. The rich diversity of mangrove trees have exposed roots, numerous branches and usually form a closed canopy. Darwin is a good place to enter the mangroves which can be hard to walk through and are best explored at low tide when the mud is exposed.

Mangroves are the primary habitat for a number of sought after species, such as the Great-billed Heron, Chestnut Rail,

Mangrove Robin, Mangrove Golden Whistler, White-breasted Whistler and Mangrove Grey Fantail. More commonly observed species include the Collared Kingfisher (along the water's edge), three species of gerygone, Brown and Red-headed Honeyeaters, Broad-billed and Shining Flycatchers, Black Butcherbird and Yellow White-eye. Mangroves have some similarities to monsoon forests and you will find some species in both habitats.

Mudflats and Beaches

Mudflats provide an important habitat for the migratory waders that arrive in and depart from Australia through the Top End. They generally occur at the mouths of mangrove-fringed tidal creeks, are unvegetated and provide a rich source of food for waders.

Beaches, sand dunes and rock platforms line parts of the coast not fringed by mangroves and mudflats. The Darwin area provides the best opportunities for exploring these habitats as most other coastal areas are inaccessible. To plan coastal birdwatching check the local paper for tide times: on mudflats, mid to high tide is the best time as the waders are forced closer to shore to feed or roost.

Waders arrive from the northern hemisphere in the early wet (October and November). The sight is spectacular, with thousands of birds arriving simultaneously. The species composition changes daily as different birds arrive, stay for a few days and then disperse throughout Australia. The departure at the end of the wet (March and April) also produces large flocks although birds tend to depart from a greater number of points. As well as the mudflats, sandy beaches provide habitat for waders, and rocky headlands provide roosts during high tides. Some birds overwinter in northern Australia and it is worth looking for waders at any time of the year. Many sightings of vagrants are recorded and birding organisations (see Appendix 1) would appreciate notification of new sightings.

Terns and gulls are often seen flying above the beaches. This book does not deal with most of the truly oceanic birds. Some species, such as the Least Frigatebird, are included as they can be seen from the coast, particularly in monsoonal conditions.

The coastal dunes and beaches support a narrow fringe of distinctive, hardy vegetation where woodland species occur — look for Yellow White-eye. Monsoon forests border some beaches and support a range of species typical of that habitat.

Wetlands

The Top End's high rainfall produces large wetlands — floodplains, billabongs and saline swamps.

The floodplains are extensive, flat and usually treeless. They become inundated in the wet when aquatic birds are spread over a vast area. As the dry season sets in, water gradually recedes to the permanent billabongs. The floodplains become covered in grasses which are inhabited by grassland species; typically, quails, button-quails, some cuckoos, Richard's Pipit and Tawny Grassbird. At this time aquatic birds gather at the permanent billabongs and swamps resulting in massive concentrations of Magpie Geese, whistling-ducks, ducks, Green Pygmy-Geese, egrets, ibis, Black-necked Stork, rails, crakes, Comb-crested Jacana and terns. In the dry, where paperbarks, Freshwater Mangroves and pandanus line the permanent watercourses, you can find Black Bittern, cuckoos, kingfishers, honeyeaters, flycatchers and orioles. Yellow Water in Kakadu National Park (Area 3) and Corroboree Billabong (Area 2) are excellent examples of this type of habitat.

Closer to the coast some areas become inundated by seawater during high tides and produce saline swamps. These are covered by low succulent bushes and sedges and in the wet season are dominated by taller grasses. They primarily house grassland birds and are the main habitat of the Zitting Cisticola.

CENTRE

Shrubland

The most widespread habitat in the Centre is shrubland, which occurs on a variety of landforms. It commonly has an overstorey dominated by Mulga, or multi-stemmed eucalypts referred to as mallee. Spinifex is a common understorey and is an important bird habitat. Scattered trees occur within a shrubland, sometimes producing woodland pockets; Ironwood, Whitewood and corkwoods are common woodland species. Corkwoods are nectar-producing species that, when in flower, attract large numbers of nectar feeders.

Common shrubland birds include cockatoos and parrots, fairy-wrens, thornbills, Singing Honeyeater, Crimson Chat, robins, Crested Bellbird, Rufous Whistler, Grey Shrike-thrush and butcherbirds. In the sandier areas Desert Oaks provide an open overstorey and are a favourite haunt of cockatoos.

Acacias, eremophilas and grevilleas are important shrubland plants and when in flower attract species such as Black, Pied and White-fronted Honeyeaters.

Chenopod shrublands are primarily composed of bluebush and saltbush with little or no understorey. White-winged Fairy-wren and Cinnamon Quail-thrush are common here and Orange Chat can be abundant at times.

Riverine

The waterways of Central Australia are dry for most of the year and dominated by River Red Gums whose tall single stemmed trunks are often more than 15 metres high. Riverine habitats may contain other eucalypts, such as the Coolibah. The understorey is usually grass with scattered shrubs and herbs.

This habitat provides the greatest bird diversity in the Centre. Thornbills are common, with the Chestnut-rumped

Thornbill feeding in flood debris and the Inland Thornbill in shrub understorey. It is also a place for Rufous Whistlers, parrots, kingfishers, honeyeaters, butcherbirds, Striated and Red-browed Pardalotes and Grey Shrike-thrush. Many of the raptors nest, roost and hunt there and it is an important breeding habitat for many hollow-nesting birds including parrots and cockatoos.

Ranges

Central Australian ranges are well represented within the park system. Composed of quartzite and sandstone ridges, they are generally infertile, badly scarred by fire and exposed to extremes of temperature. For these reasons the dominant vegetation is spinifex with Ghost Gum and Bloodwood trees scattered throughout at varying densities

Pockets of remnant vegetation occur in protected areas on south-facing slopes and within the gorges and chasms. Mulga, Hill Mulga, White Cyprus Pine, Native Fig and MacDonnell Ranges Cycad are the dominant species.

Although ranges do not have a high bird diversity, some exciting and rarely seen birds live in the spinifex and sheltered gorges. The creeks running through the gorges create permanent waterholes which attract a number of species including the Western Bowerbird. The Spinifex Pigeon, Dusky Grasswren and Painted Finch inhabit the spinifex-clad rocky ranges, usually near water. The Little Woodswallow is associated with rocky gorges and the Black-chinned Honeyeater and raptors such as the Peregrine Falcon and Australian Hobby occur along the watercourses. In the denser spinifex on the lower slopes look for Spinifexbird and Rufous-crowned Emu-wren.

Sand Dunes

The large red sand dunes of the Simpson Desert run generally north/south. The slopes are covered with spinifex and the crests are dominated by Sandhill Cane-grass, with hopbushes, chenopods and acacias. Swales between the dunes may support a typical shrubland vegetation. In areas that are periodically inundated the vegetation is dominated by Lignum, while Coolibah grows along drainage lines and river floodouts.

White-winged and Variegated Fairy-wrens are common in the sand dunes, as are Singing Honeyeater, Southern Whiteface and Cinnamon Quail-thrush. Crimson Chat, Little Button-quail and Ground Cuckoo-Shrike can be common at times. The Eyrean Grasswren is found only on the crests of sand dunes dominated by cane-grass. The Banded Whiteface often appears in the interdunal swales.

Sand dunes occur in other areas in the Centre but do not produce a widespread, distinct habitat like those in the Simpson Desert.

Plains

Grass plains and gibber plains form the main body of this habitat, which is generally treeless and poorly drained.

The grass plains of the Barkly Tablelands are dominated by a dense layer of Mitchell grass. This is the stronghold of the Flock Bronzewing, a bird solely of the grassland. The Yellow Chat occurs around reedy bores; and Letter-winged Kites are sometimes found in clumps of trees. Numbers of Oriental Pratincole occur in summer; Grey Falcon are sometimes sighted and the Pictorella Mannikin inhabits vegetation around swampy areas. Perhaps the most common bird of the Mitchell grass plains is the melodious Singing Bushlark.

The red stony plains known as the gibber plains are usually devoid of vegetation although after good rains they become covered with ephemeral herbs and grasses. This is the stronghold of Gibberbird, Inland Dotterel and Cinnamon Quail-thrush, with Bourke's Parrot occurring in the Red Mulga creeklines. In good seasons many of the birds on the gibber will breed and the Little Button-quail, Australian Pratincole, Flock Bronzewing, cuckoos, Banded Whiteface and Crimson Chat may be abundant. Emus and Australian Bustards are found on all types of plains.

Wetlands

Permanent wetlands in central Australia include artificial waterbodies such as the Alice Springs sewage ponds, and natural waterholes associated with the major rivers and rocky gorges. The effluent ponds are the only waterbodies that regularly hold a significant number of waterbirds. Permanent waterholes generally house a fluctuating number of species including Pacific Black Duck, Grey Teal, grebes, Eurasian Coot, cormorants, egrets, herons and some waders. The aquatic vegetation may house crakes, Clamorous Reed-warblers and Little Grassbirds. Wetlands often produce vagrants, such as the Freckled Duck, Blue-billed Duck, Yellow Chat, Golden-headed Cisticola and waders.

Ephemeral wetlands occur after heavy rains, with waterholes forming in creeks and rivers, and low lying claypans and interdunal swales becoming temporary swamps. It is worth finding out if any ephemeral lakes have recently filled because at this time they explode with a profusion of birdlife. You may see large numbers of Pink-eared Duck, Hardhead, Swamp Harrier, Brolga, crakes, Dusky Moorhen, Black-tailed Native-hen, Eurasian Coot, stilts and dotterels. Ephemeral lakes provide important breeding opportunities for some of these species.

AREAS

Each of the following 13 birdwatching areas has its own map. The legend below applies to all maps

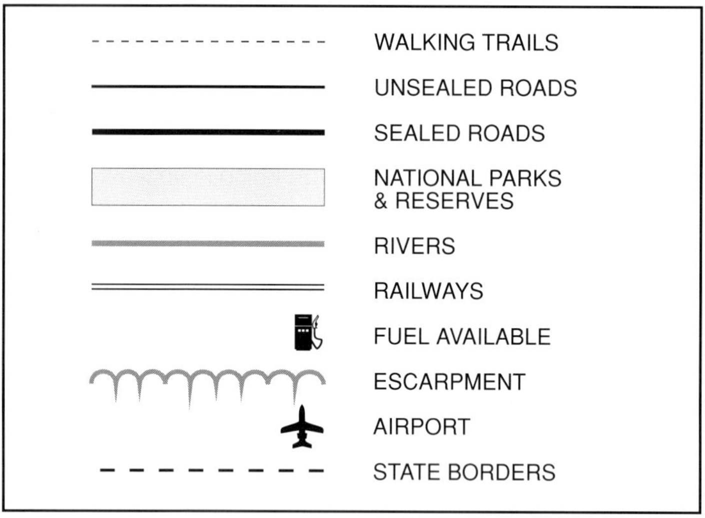

1. DARWIN

Background

Darwin provides excellent birdwatching in a wide range of habitats so it is worth spending a few days here before venturing into the bush. It is the preferred location to see certain species and is notable for vagrants.

The city and suburbs feature monsoon forest plants in the gardens and you can readily see many birds common to this habitat. Darwin is surrounded by mangroves, with good opportunities to enter them at various points. One of the best wader watching sites in the country is at Buffalo Creek.

The Checklist contains 253 species for this area.

Access

Darwin has an international and domestic airport and is accessible by bus from other major centres. A private or hire vehicle is the best way to get around the town but bicycles, public buses, taxis and guided bus tours are alternatives. Some sites are within walking distance of the city centre.

Accommodation

A full range of accommodation from backpacker hostels through to five star hotels is available. Bookings will be necessary, particularly in the dry season.

Camping

Commercial caravan parks throughout the city contain camping areas but no camping is allowed in the parks described in this chapter.

Darwin

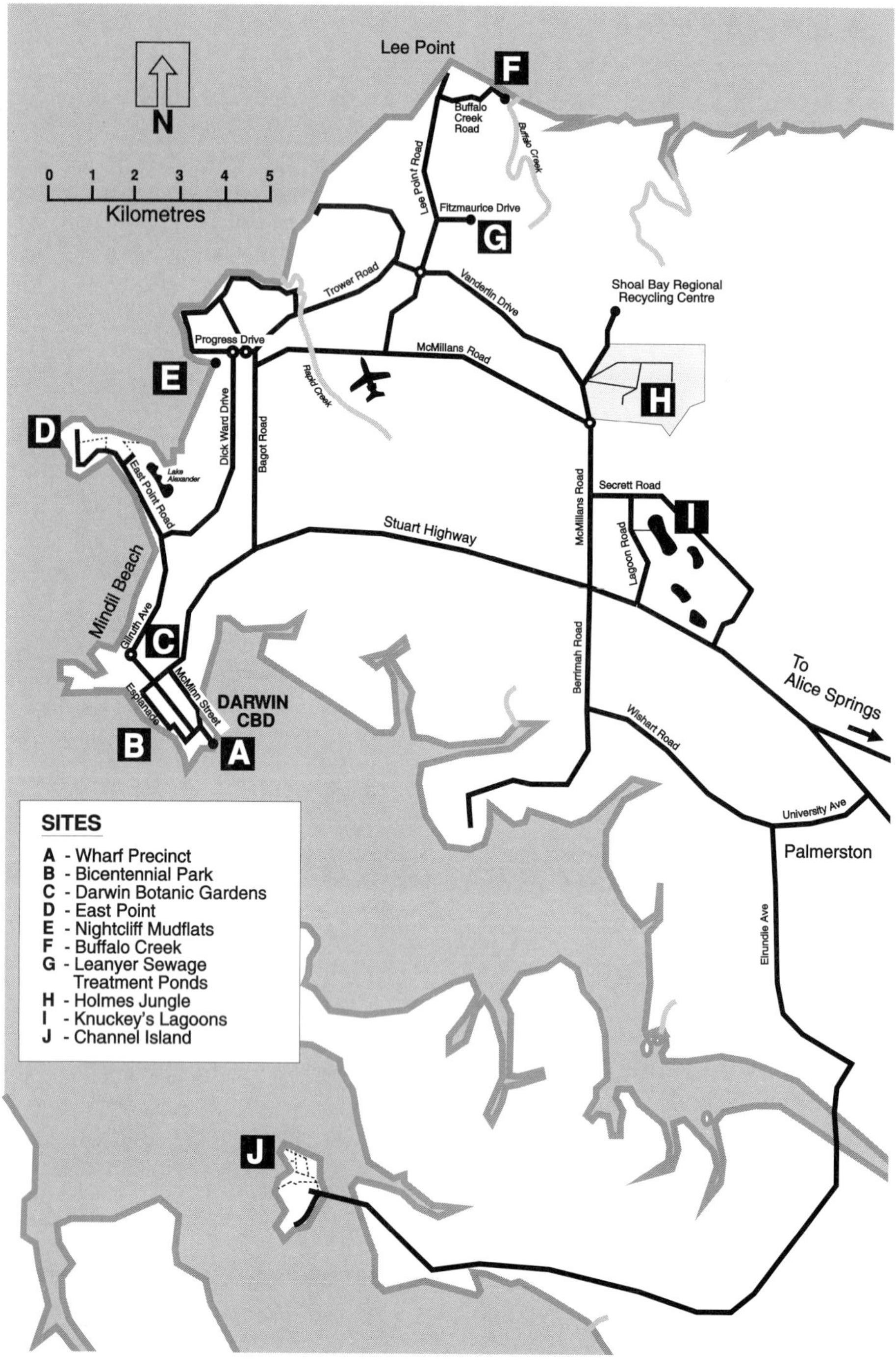

N
0 1 2 3 4 5
Kilometres
Lee Point
Buffalo Creek Road
Buffalo Creek
Lee Point Road
Fitzmaurice Drive
Trower Road
Vanderlin Drive
Shoal Bay Regional Recycling Centre
Progress Drive
McMillans Road
Rapid Creek
Dick Ward Drive
Bagot Road
Lake Alexander
East Point Road
Stuart Highway
Secrett Road
Lagoon Road
Mindil Beach
Gilruth Ave
McMinn Street
Esplanade
DARWIN CBD
Berrimah Road
Wishart Road
To Alice Springs
University Ave
Palmerston
Elrundie Ave
SITES
A - Wharf Precinct
B - Bicentennial Park
C - Darwin Botanic Gardens
D - East Point
E - Nightcliff Mudflats
F - Buffalo Creek
G - Leanyer Sewage Treatment Ponds
H - Holmes Jungle
I - Knuckey's Lagoons
J - Channel Island

Special Birds

	Recommended Sites									
	A	B	C	D	E	F	G	H	I	J
King Quail							•	•	•	
Garganey							•		•	
Great-billed Heron						•				
Chestnut Rail						•				•
Red-backed Button-Quail								•		
Red-chested Button-quail								•		
Little Curlew			•					•	•	
Broad-billed Sandpiper						•	•			
Beach Stone-curlew				•		•				•
Varied Lorikeet						•		•		
Large-tailed Nightjar				•		•		•		
Little Kingfisher						•				
Collared Kingfisher	•			•	•	•				•
Green-backed Gerygone		•	•	•		•		•		
Helmeted Friarbird		•		•	•	•				•
Mangrove Robin						•				•
Mangrove Golden Whistler				•		•				•
White-breasted Whistler						•				•
Broad-billed Flycatcher				•		•				•
Mangrove Grey Fantail						•				•
Zitting Cisticola							•	•		

Sites

The description of sites begins at the city centre and follows a roughly clockwise direction around the suburbs. The map shows the location of the sites and a Darwin street map is also recommended.

A. Wharf Precinct

On the south-eastern end of the city block the Wharf Precinct is a short walk from the city. Follow McMinn Street on to Stokes Hill Wharf where parking is permitted. A number of harbour boat tours leave from the wharf precinct.

Gull-billed, Crested, and Whiskered Terns are common along the wharf and in the Wet you may also see Caspian, Lesser-crested, Common, Little and White-winged Black Terns. The terns can be seen flying alongside the wharf or perched on buoys, small boats and pylons.

Check the ladders, scaffolding and moored boats at the beginning of the wharf for Eastern Reef Egret, Striated Heron and Nankeen Night Heron. At low tide a mudflat becomes exposed inside the wharf which is worth a quick look for egrets and waders. Collared Kingfishers roost under the wharf and feed on the mudflats at low tide.

B. Bicentennial Park

Running alongside the Esplanade on the south-western side of the city block, this park provides a pleasant walk at sunset. It is a good starting point for birdwatchers new to the Territory because of the large number of common birds found here. These include doves, Pied Imperial-Pigeon, Sacred Kingfisher, honeyeaters, cuckoo-shrikes, Varied Triller, orioles and Figbirds.

From the paths climbing down through the monsoon forest to the beach you may see gerygones and, flying above the canopy at dusk, Australian Hobbys. The beach sometimes has a few wader species. Common Koel can be found in the denser foliage during the wet season. Many birds are attracted to the well watered lawns, almost continuous flowering trees and shrubs, and the profusion of fruit-bearing palms.

C. *Darwin Botanic Gardens*

The Darwin Botanic Gardens, located a few kilometres from the city centre, provide interesting birdwatching. Enter from Gardens Road (off Gilruth Avenue). They contain monsoon forest, woodland, an artificial lake and extend to a small coastal area just north of Mindil Beach.

The gardens include a variety of introduced and native plants that flower and fruit throughout the year and attract fruit and nectar eating birds, including Rose-crowned Fruit-Dove and Pied Imperial-Pigeon. The Gardens Road entrance is a good place to find some of the gerygones, and the open lawns are good for Magpie Goose, Radjah Shelduck, Glossy Ibis, Australian White Ibis and, in the late dry season, Little Curlew. A wetland system is being developed which should attract waders and other waterbirds.

The nearby golf course has an artificial lake with some waterbirds. Woodland dwellers such as Sacred Kingfisher, White-breasted Woodswallow and martins, inhabit the more open areas. The section of coast from Mindil Beach to East Point is a haunt for Lesser Frigatebird during monsoonal activity, as is the area off Lee Point.

D. *East Point*

Located at the end of East Point Road, this reserve contains beaches, tidal flats, woodland, mangroves and monsoon forest. Walking paths, bike tracks, picnic grounds and barbecues are provided.

A mangrove boardwalk starts at the car park on the northern end of Lake Alexander (where you can swim all year) and here you can find the more common mangrove species, such as Red-headed Honeyeater, Lemon-bellied Flycatcher and occasionally Mangrove Golden Whistler.

The monsoon forest walk is on the right as you drive from the lake out to the point — look for the break in the forest. Watch for Rose-crowned Fruit-Dove and Rainbow Pitta as well as other monsoon forest species. Near the Military Museum, explore mudflats and rock platforms around the point where wader species roost on the incoming tides and the Beach Stone-curlew lives. Because the mudflats here are too soft to walk on, those at Nightcliff and Buffalo Creek are better for wader watching.

E. Nightcliff Mudflats

The Nightcliff mudflats are a good place to see a number of migratory and overwintering waders as walking here is comparatively easy. Enter opposite the intersection of Hickory Street and Progress Drive in Nightcliff. The mudflats, fringed by mangroves and exposed at tides of five and a half metres and below, can be very hot. Most of the waders feed on the water's edge which at low tide can be more than a kilometre out, so to save yourself a long, hot walk, arrive when the tide is about five metres and falling.

Good numbers of waders are present all year round, but the best time is between September and December when most of the passage migrants arrive. A number of waders, such as godwits, Eastern Curlew and plovers, usually overwinter here. Red-capped Plover is resident and you will occasionally see Osprey and White-bellied Sea-Eagles.

The rocky point to the right is good for Ruddy Turnstone and Sooty Oystercatcher. The mudflats and mangroves are good places to find mangrove species including Collared Kingfisher, Mangrove Gerygone, Helmeted Friarbird and Yellow White-eye.

F. Buffalo Creek

Drive to the carpark at the end of Buffalo Creek Road where the four main habitats of mudflats, monsoon forest, mangroves and woodland are easily accessible. This is Commonwealth land managed by the Royal Australian Air Force. The places described are open to the public except for one day a year when a road block is set up in order to meet some legislative requirements.

This beach and mudflat area is the most accessible wader watching site in the Northern Territory and one of the best as it is an arrival and departure point for the migrants. Waders concentrate on the sand spit at the mouth of the creek and on the beach in front of the monsoon forest. It is not uncommon to see flocks of thousands here from October to March and at least some overwintering waders are present all year. Great Knots are usually the most numerous all year round as they do not disperse from this area to the same extent as other waders. Different species of migratory waders arrive at different times and one species may become very common or predominant for a short time, for example Curlew Sandpiper can be seen in flocks of hundreds for a week or two. Almost all Australian coastal waders can be seen here including Terek Sandpiper and Sanderling which are usually present in small numbers; you will also find Red Knot, which is difficult to separate from Great Knot. This is a good place for Broad-billed Sandpiper which frequents the muddier areas such as the mouth of Buffalo Creek. Some waders pass through as passage migrants on the return trip during March and April, which can also be a good time for wader watching.

The site is well-known for vagrants (we once saw a Flock Bronzewing knee deep in sea water here).

Raptors are also common; White-bellied Sea-Eagle, Australian Hobby and Peregrine Falcon are often seen patrolling the beach.

Monsoon forest lines both sides of the last section of road before the carpark. You can walk along the road or take the short tracks from the road to the beach through this habitat. The place is heavily used, especially on weekends, so shyer birds such as Rainbow Pitta probably won't be seen. However, there are usually good numbers of birds here including Orange-footed Scrubfowl, Emerald Dove, Little Bronze-Cuckoo, Grey Whistler, Little Shrike-thrush, Northern Fantail and Yellow Oriole and, from September to November, Oriental Cuckoo, Common Koel and Dollarbird. Green-backed Gerygone and Yellow White-eye are present all year and often perch in vine tangles and in broad-leaved trees along the beach edge. Also watch for Helmeted Friarbird and Black Butcherbird which can be seen in the forest or perched on dead branches above the canopy. Honeyeaters can be abundant along the road, with sometimes up to eight species present, and Varied Lorikeets arrive in numbers in the dry when the paperbarks are in flower. Large-tailed Nightjar is resident here and can be seen at dusk over the grassy area near the 90 degree bend in the road that marks the start of the monsoon forest.

The Buffalo Creek site also contains extensive mangroves. Birds such as Red-headed Honeyeater, can be seen from the carpark, or you can walk a short way up the creek from the boat ramp. Azure and Collared Kingfishers, Mangrove and Large-billed Gerygones, Shining Flycatcher and Mangrove Grey Fantail are all likely along this short stretch.

If you really want to get into the mangrove forest proper you can continue upstream but it is hard work and crocodiles may be lurking: it is much easier and safer by boat. If you are prepared and keen you will explore a good mangrove forest and have a chance of seeing Great-billed Heron, Chestnut Rail (in the forest or on the mudbanks at low tide), Little Kingfisher, Mangrove Golden Whistler, Broad-billed Flycatcher and the other mangrove species mentioned above. The White-breasted Whistler has been seen here in dense stands of stunted mangroves and the Mangrove Robin is also present. A short path starts at the corner of the carpark (furthest from the boat ramp) which is an easy walk through mangrove and monsoon forest for about half a kilometre to the main road. This is a good option if you do not want to exert yourself too much. The woodland lining the road from the turnoff to the start of the monsoon forest has good numbers of birds that are generally common in the Top End.

G. Leanyer Sewage Treatment Ponds

The sewage treatment ponds located at the end of Fitzmaurice Drive have been a haunt for local birdwatchers for years because of their close proximity to Darwin, abundant birdlife and ability to produce rarities. However, access is restricted and obtaining a key to the gate may be difficult. Contact the Northern Territory Field Naturalists or the Power and Water Authority for advice on access.

The ponds farthest from the gate usually contain many waterfowl and during the wet season Garganeys have been seen. Other rarities have turned up, notably Yellow and White Wagtails. White-bellied Sea-Eagles are resident and often cruise the ponds, flushing most of the waterfowl into the air.

Numerous tracks and rough roads continue outside the gate. Tracks to the south of the fenced effluent ponds pass through vacant land with woodland, samphire, mangroves, grassland and a small patch of monsoon forest where birds from these habitat types can be seen. Watch for King Quail in the grassland. Large numbers of Chestnut-breasted Mannikins are present in the wet and you may see Star Finches and Yellow-rumped Mannikins. In the dry, watch for Long-tailed and Masked Finches. Zitting Cisticola (hard to separate from the more numerous Golden-headed Cisticola) has been seen in the saline swamp past the pipeline but access is difficult.

H. Holmes Jungle

This is the best monsoon forest site near Darwin. Access to this park is off Vanderlin Drive. Internal roads are unsealed and may become tricky in the Wet. The scenic lookout at the upper carpark provides views of monsoon forest, ridges, woodland and wetland and the lawn areas surrounded by long grass are excellent places to look for finches. Seven species have been recorded here including Star Finch and Yellow-rumped Mannikin.

The park includes a large patch of monsoon forest with walking paths where you can see a good variety of monsoon forest birds, including Rose-crowned Fruit-Dove, Rainbow Pitta and Rufous Fantail. The tracks generally follow Palm Creek which is a good place to look for the Azure Kingfisher. Birds are often sparse in the forest and the edges can be better; Grey Whistler is common here. The woodland area holds many species including Forest Kingfisher, Rainbow Bee-eater in the dry and Dollarbird in the wet. The ephemeral wetlands adjacent to the forest sometimes hold many waterfowl as well as crakes. Brown and King Quails and Red-backed and Red-chested Button-quails occur on the floodplains. Zitting Cisticola is conspicuous at the beginning of the wet season as it constantly calls and performs display flights.

I. Knuckey's Lagoons

To find Knuckey's Lagoons, turn off McMillans Road into Secrett Road, then right into Lagoon Road from which the main lagoon is visible on the left. At the first bend in Lagoon Road take the grassy track between two fenced paddocks which leads to the lagoon. This is a good site for observing waterbirds. From September to December, you can expect to see some waders, particularly the Little Curlew, Sharp-tailed Sandpiper and the Curlew Sandpiper. The Black-necked Stork, Comb-crested Jacana and four species of egret all live here. Gull-billed and Whiskered

Terns are common and the White-winged Black Tern is present in good numbers from October to March. Large flocks of Pied Herons arrive to roost at sunset. In the surrounding grassland you can find snipe, Singing Bushlark and possibly Brolga.

J. Channel Island

Channel Island houses the power station that supplies Darwin's electricity. To get there, from Palmerston turn off University Avenue into Elrundie Road and follow it to the end. This road travels through an excellent area of woodland which is worth looking at. On the left, just before the island an area of stunted mangroves and other salt tolerant vegetation may hold Mangrove Robin and White-breasted Whistler.

Once on the island, enter the walking track from a gate on the right opposite the boat ramp turnoff. The power station entrance building has brochures about the walking track and a key is available if the gate is locked. On the track, the first turn to the left leads uphill and then to the old leprosarium site. Don't take this turn, but walk on ahead to the mangrove area.

The site has excellent access to a rich and diverse mangrove habitat and nearby monsoon forest. This is the most accessible mangrove site in the Territory and all the mangrove frequenting species can be found here. The trail has an extensive boardwalk section giving excellent observation over the canopy of mangroves. It passes the inland edge of the mangroves where there are opportunities to see Mangrove Robin and White-breasted Whistler. The Mangrove Gerygone, Helmeted Friarbird, Red-headed Honeyeater and Lemon-bellied Flycatcher are common. In the monsoon forest, you can expect to see Rainbow Pitta and other typical forest species.

2. SMALL PARKS OF THE TOP END

Background

The sites in this area provide excellent opportunities for viewing a wide variety of the Top End's birds. Fogg Dam in particular has an astounding array of waterbirds and is also one of the best places in the Top End to view monsoon forest. Howard Springs and Berry Springs also have good examples of monsoon forest along with popular swimming holes. Corroboree Billabong has classic Top End floodplain and associated habitats.

The Checklist contains 220 species for this area.

Access

Fogg Dam, Howard Springs and Berry Springs are easy day trips from Darwin while Corroboree Billabong probably requires an overnight stay. All roads except the one to Corroboree Billabong are sealed and open all year. The road to Corroboree Billabong is unsealed and may be impassable in the wet.

Accommodation

In addition to facilities provided in Darwin, accommodation is available at Humpty Doo, Corroboree Park, Annaburroo, Wildman Reserve, Noonamah and Berry Springs.

Camping

Park camping is only allowed at Wildman Reserve (next to Corroboree Billabong) and at Corroboree Billabong. Commercial caravan and camping facilities are provided at the locations listed above under accomodation.

Small Parks of the Top End

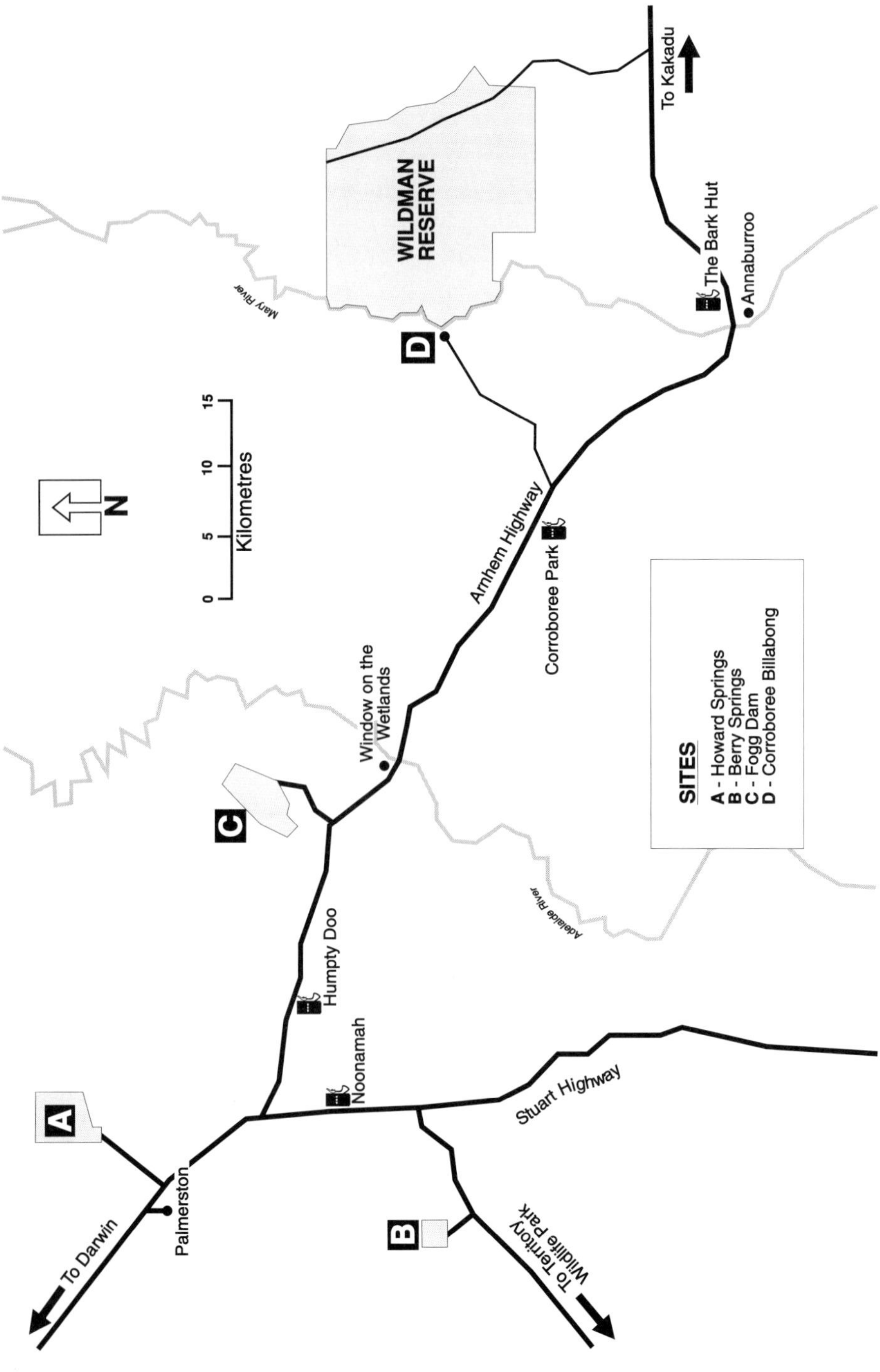

Special Birds

	Recommended Sites			
	A	B	C	D
Garganey			•	•
Black Bittern				•
Letter-winged Kite			•	
White-browed Crake			•	
Red-chested Button-quail			•	•
Little Curlew			•	
Varied Lorikeet	•	•	•	•
Oriental Cuckoo	•	•	•	•
Rufous Owl		•	•	
Large-tailed Nightjar			•	•
Rainbow Pitta	•		•	
Bar-breasted Honeyeater			•	
Broad-billed Flycatcher			•	

Sites

A. Howard Springs

To reach the Park from Darwin, take the Stuart Highway and turn left on to the Howard Springs Road. The park is open from 8 am to 8 pm and contains picnic areas, barbecues and a kiosk. The large spring-fed swimming and paddling pools are surrounded by monsoon forest. Watch for large barramundi and other fish in the main pool behind the weir —no fishing allowed.

The signposted walking track below the weir passes mainly through monsoon forest and allows you to walk silently while viewing the canopy. You can easily see Rose-crowned Fruit-Dove, Pied Imperial-Pigeon, Rainbow Pitta, Shining Flycatcher, Rufous Fantail and orioles. Watch for Azure Kingfisher over the flowing creek.

Another short walking track (off the road to the eastern carpark) goes through monsoon forest to the springs which feed the main pool.

The woodland fringing the entry road has grevilleas and other trees which attract numerous species of honeyeaters (notably Bar-breasted) when in flower. Lorikeets can also be common.

The picnic areas attract many birds. Orange-footed Scrubfowl, Australian White Ibis, Straw-necked Ibis, Tawny Frogmouth, Blue-winged Kookaburra and White-gaped Honeyeater are common and tame. Nankeen Night Heron is tame at the cement overflow to the springs and is often active during the day.

B. Berry Springs

This park provides picnic areas and swimming in freshwater springs and is open from 8 am to 6.30 pm. Follow the signs on the Stuart Highway to the nearby Territory Wildlife Park (which is also worth a visit).

Berry Springs is very similar to Howard Springs from a birdwatching perspective. The picnic area has a number of easily observed tame species and the short circuit walk passes through a small patch of monsoon forest. At the beginning of this walk flowering grevilleas and eucalypts can be a good spot for lorikeets and honeyeaters. Northern Rosella is often observed in the woodland.

A walking trail with a boarded section crosses the creek. Watch for Rufous Owl during the day roosting near the fig tree and just before dark when they become active. The fruit of the large fig tree attracts fruit eaters including Rose-crowned Fruit-Dove, Pied Imperial-Pigeon, lorikeets and orioles. Large-billed Gerygone sometimes perch in the overhanging vegetation at the swimming hole and Azure Kingfisher is common.

C. Fogg Dam

The internationally known Fogg Dam on the Arnhem Highway is a birdwatcher's paradise which has open water, floodplain, monsoon forest and swamp all within easy walking distance. The spectacular views over the water at sunset are popular with photographers. Plan to stay at least three hours to appreciate the birdlife and take mosquito repellent.

The dam was built in the 1950s to irrigate rice crops on the Adelaide River floodplain. When the venture folded, the dam became a wildlife sanctuary.

The best time to visit the reserve is in the late dry season when waterfowl flock here as other waterbodies become dry. However, the dam is definitely worth a visit at any time of year. Early mornings are the best time to go.

The dam wall has three viewing platforms and is excellent for watching waterbirds. All waterbirds of the Top End are regularly spotted here. On the open water you can easily see Darter, cormorants and other waterfowl. Garganey has been recorded during the wet season and you may find the Little Curlew. Watch for Baillon's Crake, White-browed Crake and Comb-crested Jacana in the lilies. Double-barred Finch, Crimson Finch, Chestnut-breasted Mannikin, Tawny Grassbird and Golden-headed Cisticola are common in the pandanus and grasses on the side of the dam wall, where you may also see Yellow-rumped Mannikin.

The downstream side of the dam is dominated by grassland in the dry season and is inundated in the wet. During the dry, you can see Black-necked Stork, Brolga, Purple Swamphen and Australian Pratincole. The Little Corella feeds in huge flocks on the plains. At the end of the wet and into the early dry the spillway at the other end of the dam wall houses the Little Egret and at sunset, large flocks of Nankeen Night Heron appear.

The White-winged Black Tern is a summer migrant to inland waterways close to the coast, which can be found at Fogg Dam, at times in flocks of up to a couple of hundred birds. In Australia, it is mostly seen in non-breeding plumage when it can be difficult to separate from the Whiskered Tern. In breeding plumage it is a spectacular bird.

In the paperbarks fringing the carpark end of the dam wall, watch for flycatchers including the Broad-billed, Leaden, Restless and, closer to the water, Shining Flycatcher. The Rose-crowned Fruit-Dove is easily observed and you may also see the Bar-breasted Honeyeater, Lemon-bellied Flycatcher and Grey Whistler.

Next to the carpark is a signposted walk through the monsoon forest which can be impassable in flood conditions. When birdwatching here, stay on the path — it is difficult to walk quietly in the forest because of the leaf litter. The crackling of dry leaves underfoot scares the birds and makes it difficult to locate them by call. This walk provides an excellent opportunity to observe monsoon forest species such as the Rainbow Pitta. The largest known population of this bird is found here and it is often easy to observe. The pittas are found in the darker forest where the understorey thins out and leaf litter predominates. You may detect them by their shuffling in the leaves. Orange-footed Scrubfowl is common and often found in the same way. The Rose-crowned Fruit-Dove is also common but difficult to observe in the canopy. It is best located by its distinctive call and from falling fruit as it feeds. Other birds to look for in the forest include: the Pacific Baza, Oriental Cuckoo, Brush Cuckoo, Little Bronze-Cuckoo, Dusky Honeyeater, Lemon-bellied Flycatcher, Broad-billed Flycatcher, Leaden Flycatcher and Rufous Fantail. Watch again for the Broad-billed Flycatcher, Shining Flycatcher and possibly Barking Owl at the end of the boardwalk. Brolga can be observed on the nearby floodplain.

Night time birdwatching can be very rewarding on the dam wall. There is a good chance of seeing the Bush Stone-curlew and Southern Boobook, and closer to the monsoon forest you may see the Barking Owl, Large-tailed Nightjar and Rufous Owl. Ranger-guided night walks take place in the dry season.

The Letter-winged Kite has inhabited the eucalypt vegetation at the intersection of the Fogg Dam road and the Arnhem highway in recent years. At dusk these birds become active and feed in the short grassland on the nearby Royal Australian Navy property. They have been seen here on antennae and in the solitary eucalypts in the cleared paddock. The species is highly nomadic so there is no guarantee that this population will stay. Finches and Tawny Grassbird are common in the area.

Just past the Fogg Dam turnoff on the Arnhem Highway, extensive wetlands fringe the road. During the dry season these contain numerous waterbirds. This site is grazed by domestic buffalo so the short pasture may offer different habitat or better viewing for some species possibly not seen at Fogg Dam. It is worth a stop even during the wet season.

At the end of the wet, water drains off the wetlands through a road culvert at the east end of the wetland attracting numerous egrets, herons and terns.

The Arnhem Highway itself offers great birdwatching, particularly for raptors. The Adelaide River floodplain houses numerous raptors, especially in the dry season. The Square-tailed Kite is found in the nearby wooded country and the Black-breasted Buzzard is not uncommon.

The Window on the Wetlands Visitor Centre just past this area is worth a stop for an introduction to the ecology of the wetlands.

D. Corroboree Billabong

Although it is just outside the Wildman Reserve, public access is allowed to the boat ramp on this billabong, which forms part of the Mary River wetlands. It is about an hour and a half drive from Darwin and is definitely worth a visit during the dry season. Take the Mary River Houseboats road to the north of the Arnhem Highway a few kilometres past Corroboree Park. Follow the signs to Mary River Houseboats and continue along the road to the boat ramp.

During the dry season, aquatic and terrestrial birdlife is abundant. Corroboree Billabong contains numerous Estuarine Crocodiles and mosquitoes and other insects make camping extremely trying if you are not well prepared. At night, during the dry season, it can be one of the cooler places in the Top End.

The profusion of waterbirds are easy to observe on the short grassed floodplains. Glossy Ibis and Brolga usually occur in good numbers. Horsfield's Bronze-Cuckoo and Tawny Grassbird are common and you may see the Black-eared Cuckoo and Red-chested Button-quail. During the wet or early dry season look for Channel-billed Cuckoo near the fig trees on the water's edge, a couple of hundred metres upstream of the boat ramp. Bush Stone-curlew roost in the clumps of taller grasses and shrubs on the edge of the floodplain. The water's edge is dominated by pandanus, Freshwater Mangroves and paperbarks, which are inhabited by Nankeen Night Heron, Black Bittern, Azure Kingfisher, Little Kingfisher and Shining Flycatcher. When in flower, the fringing paperbarks attract honeyeaters and Varied Lorikeet.

Nocturnal birds are common. On the edge of the waterway, particularly near the boat ramp, the Barking Owl becomes active at dusk and is easily located by its distinctive call. Another way to see it is to drive along the dirt road leading in from the highway just after dusk. The Spotted Nightjar is common and you can see the Large-tailed Nightjar near monsoon forest on the edge of the floodplain. You can also usually see the Bush Stone-curlew, Southern Boobook, Barn Owl and Australian Owlet-Nightjar.

3. KAKADU

Background

Featuring the spectacular scenery of the Arnhem Land escarpment, Kakadu National Park includes four major river systems and wetlands of international significance.

The Checklist contains 288 species for this area, which is superb for birdwatching and includes all four Territory endemics.

Access

From Darwin, follow the Arnhem Highway; from the south, use the Kakadu Highway. Access to and within the park can be severely restricted during the wet. Check with the park headquarters for information. The entry fee, which covers a fortnight, can be paid at either entrance station or at the information centre at park headquarters. With your entrance ticket you receive a Visitor Guide that includes a detailed map of the park and its facilities. All sites mentioned in this chapter are on the map. If you are planning several visits, it may be worth purchasing a Territorian Park Use ticket which lasts for one year and includes access to Uluru-Kata Tjuta National Park.

Accommodation

A range of accommodation from hotel through to backpacker standard is provided in the town of Jabiru, at Cooinda and at Kakadu Holiday Village.

Camping

Jabiru, Cooinda, Kakadu Holiday Village and Mary River Roadhouse have commercial campgrounds and caravan parks. Mardugal, Muirella Park, Merl and Gunlom Park all have camping areas with caravan and shower facilities. Campgrounds with more basic facilities are free and are found at a range of locations marked on the Visitor Guide map.

Kakadu

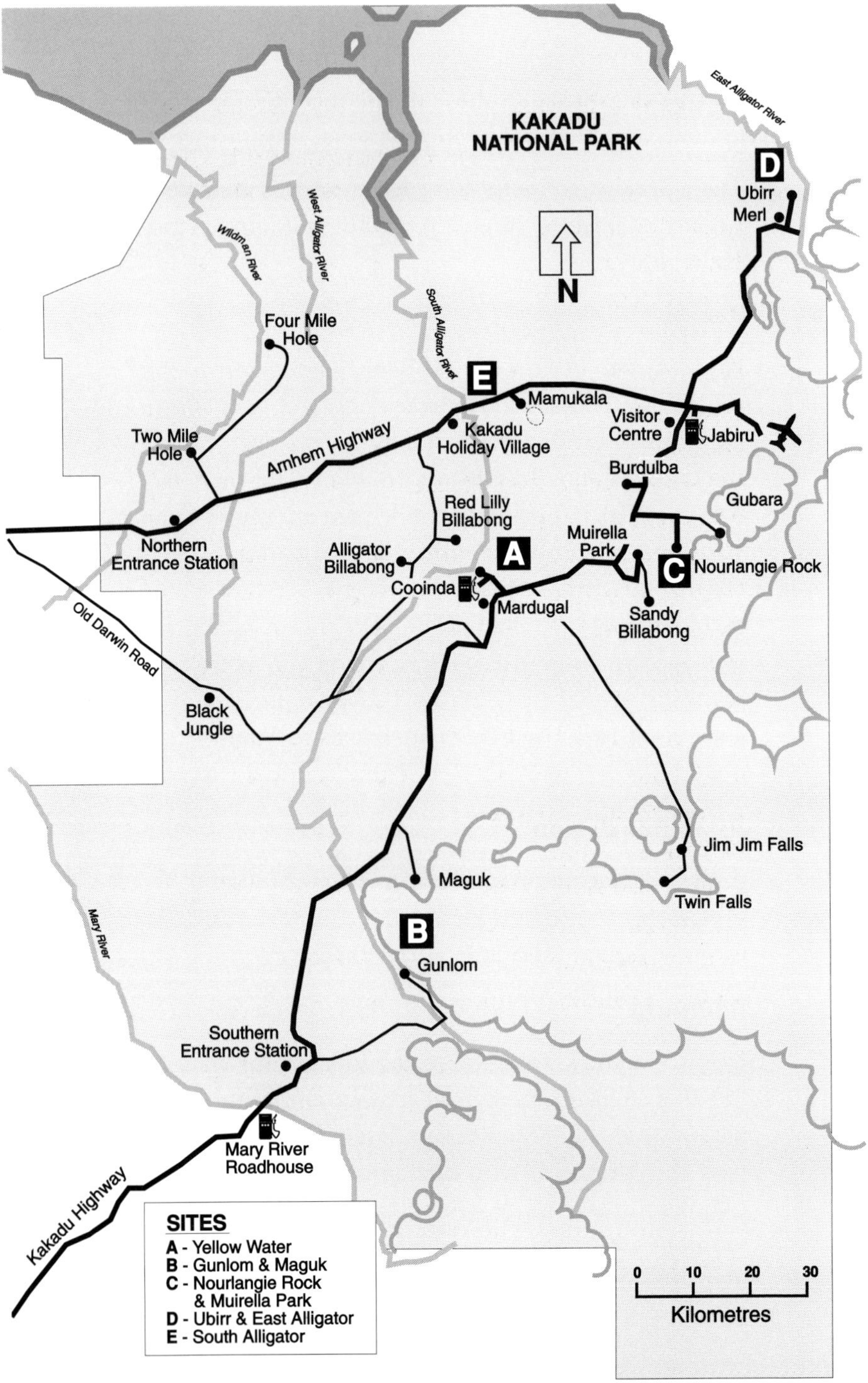
KAKADU NATIONAL PARK
N
East Alligator River
West Alligator River
Wildman River
South Alligator River
Mary River
Ubirr
Merl
Four Mile Hole
Two Mile Hole
Arnhem Highway
Mamukala
Kakadu Holiday Village
Visitor Centre
Jabiru
Burdulba
Gubara
Northern Entrance Station
Red Lilly Billabong
Alligator Billabong
Cooinda
Mardugal
Muirella Park
Nourlangie Rock
Sandy Billabong
Old Darwin Road
Black Jungle
Jim Jim Falls
Twin Falls
Maguk
Gunlom
Southern Entrance Station
Mary River Roadhouse
Kakadu Highway
A
B
C
D
E
SITES
A - Yellow Water
B - Gunlom & Maguk
C - Nourlangie Rock & Muirella Park
D - Ubirr & East Alligator
E - South Alligator
0
10
20
30
Kilometres

Special Birds

	Recommended Sites				
	A	B	C	D	E
Great-billed Heron	•				
Black Bittern	•	•			•
Red Goshawk		•			
Red-backed Button-quail				•	
Chestnut-backed Button-quail				•	
Little Curlew					•
Partridge Pigeon	•		•	•	
Chestnut-quilled Rock-Pigeon		•	•		
Banded Fruit-Dove		•	•		
Varied Lorikeet	•	•		•	•
Northern Rosella	•			•	
Oriental Cuckoo	•				
Rufous Owl	•				•
Little Kingfisher	•				•
Rainbow Pitta		•	•	•	
Black-tailed Treecreeper	•			•	
White-throated Grasswren		•			
Helmeted Friarbird		•	•		
White-lined Honeyeater		•	•		
Yellow Chat					•
White-browed Robin		•			
Sandstone Shrike-thrush		•	•		
Zitting Cisticola					•

Sites

This book describes the best birdwatching sites but Kakadu has many other places of interest — use your Visitor Guide map and go exploring.

A. Yellow Water

The most famous wetland in Kakadu, Yellow Water retains a large body of water throughout the dry. It is undoubtedly one of the best places in Australia to view waterbirds and visitor numbers are high. The best way to experience the wetlands is by boat. Tours can be booked at Cooinda but make sure you book in advance. A front seat on a dawn tour is the prime position.

All the birds in the Territory associated with wetland habitats come to Yellow Water. Magpie Goose, Plumed Whistling Duck and Wandering Whistling Duck occur in enormous numbers. Green Pygmy-Goose, Darter, cormorants, Australian Pelican, egrets, ibis, spoonbills, Black-necked Stork, Whistling Kite, Brolga and Comb-crested Jacana are also common.

A big advantage of the boat trip is that you are close to vegetation along the banks which is inhabited by interesting but otherwise difficult to locate birds such as Nankeen Night Heron and Black Bittern. People regularly see the Great-billed Heron in the Freshwater Mangroves and along tributaries. Azure and Little Kingfishers are shy and unobtrusive and are seen perched over the water — scan the vegetation and look for flashes of blue. You will often see the Barking Owl roosting in the mangroves and the Tawny Frogmouth camouflaged in forks of trees. You may also spot the Rufous Owl and a good variety of other birds, such as White-winged Black Tern, pigeons, cuckoos, Forest and Sacred Kingfishers, honeyeaters, flycatchers and trillers.

Birdwatchers regularly see Osprey, White-bellied Sea-Eagle, Grey Goshawk and other raptors, often on exposed perches and obviously taking advantage of the abundant wildlife. Thousands of Little Corella feed on the floodplain.

On the Yellow Water walk from Cooinda to the boat ramp you should see Tawny Grassbird and Golden-headed Cisticola in the taller clumps of grasses, and cuckoos in the scattered trees. Around the carpark watch for Lemon-bellied, Leaden and Restless Flycatchers and other woodland birds. In the wet, people have seen Oriental Cuckoo in the area. You may see Barking Owl and Spotted Nightjar around the lights at the Cooinda resort and the Bush Stone-curlew is a regular visitor.

Mardugal campground has an excellent woodland walk with good chances of spotting Partridge Pigeon, Northern Rosella and Black-tailed Treecreeper. When the Woollybutt is in flower, Varied Lorikeet and Bar-breasted honeyeater are common.

B. Gunlom and Maguk

Also known as Waterfall Creek or UDP Falls, Gunlom is the best site to look for Chestnut-quilled Rock-Pigeon, Banded Fruit-Dove and White-throated Grasswren, which are all Territory endemics. Variegated (Lavender-flanked) Fairy-wren, Helmeted (Sandstone) Friarbird, White-lined Honeyeater and Sandstone Shrike-thrush also occur here.

To climb the escarpment, follow the track to the right of the waterfall. The White-throated Grasswren keeps to spinifex dominated areas at the top. You may have to cover some territory as it can appear anywhere in the right habitat. The Chestnut-quilled Rock-Pigeon is a common bird which you should find reasonably easily on the rocky edges. Also watch for Sandstone Shrike-thrush calling from the top of rocky outcrops. Look for Variegated Fairy-wren in the scrubby vegetation on the slopes of the escarpment and Red-backed Fairy-wren on the escarpment. The Helmeted Friarbird and White-lined Honeyeater are found along the watercourse on top. When the Woollybutt trees are covered with their orange blossoms (June–July) these honeyeaters use the nearby woodland vegetation, and other honeyeaters and lorikeets appear in abundance.

The Banded Fruit-Dove uses patches of monsoon forest both on top and at the base of the escarpment. This bird is difficult to find in the forest but when flying between forest patches it may perch in the Cypress Pines on the escarpment. The easiest place to observe the monsoon forest is around the waterhole at the base of the falls. Look for other monsoon forest species such as Rose-crowned Fruit-Dove, White-browed Robin and Shining Flycatcher. Also, look for Red Goshawks in this area.

Maguk, or Barramundi Gorge, is designated as four-wheel drive access. It has a walking track leading through monsoon forest to an excellent swimming hole. You usually see Nankeen Night Heron, Emerald Dove, Rainbow Pitta and Grey Whistler. Look for Black Bittern in the pandanus around the waterhole.

Gunlom and Maguk are often closed in the wet when Nourlangie Rock provides a good alternative for some of the escarpment and monsoon forest species.

C. Nourlangie Rock and Muirella Park

Nourlangie Rock is an excellent area for birdwatching and provides good access to the escarpment. The Angbangbang Rock Shelter walk takes you to one of the two prime rock art sites in Kakadu. You can expect to see the more common escarpment species such as Chestnut-quilled Rock-Pigeon, Helmeted Friarbird, White-lined Honeyeater and Sandstone Shrike-thrush here, but not the White-throated Grasswren. The Broad-billed Flycatcher is found in the creekline. The extended Barrk Sandstone walk over the rock to Nanguluwur Gallery is strenuous but provides further opportunities to see the escarpment species. Also look for Australian Hobby and Peregrine Falcon.

Close by, the Angbangbang Billabong Walk provides an opportunity to view wetland species and goes part way up an outlying rock where it may be worth looking for escarpment species. The Gubara Walk is at the end of a four-wheel drive track to the north of Nourlangie Rock and leads through woodland to a good patch of monsoon forest. Here you might see Banded Fruit-Dove and Rainbow Pitta. Access to the escarpment is also possible here.

The nearby Muirella Park campground is located in woodland near a billabong. You can see common woodland birds in the area and the fringes of the billabong produce a wide variety of species. The Bubba Walk will take you through a paperbark area; watch for White-throated Gerygone and other common birds.

Sandy Billabong is a more secluded area and is an excellent place to look for waterbirds, especially late in the dry. Although not in spectacular numbers, a huge variety of birds are present here.

D. Ubirr and East Alligator

Merl campground services Kakadu's other prime rock art site, at Ubirr. The campground is surrounded by woodland which is well worth exploring for a good variety of common birds including Great Bowerbird and the more sought after Partridge Pigeon, Varied Lorikeet, Northern Rosella, Black-tailed Treecreeper and Varied Sitella. Brown Quail is common and you may see Red-backed and Red-chested Button-quail. The Red-backed is supposed to be the most common of these, especially in the wet. The top of Ubirr offers a fantastic view of the floodplains and possible sightings of Magpie Goose, Black-necked Stork and Brolga.

The Manngarre Forest walk explores a patch of monsoon forest on the banks of the East Alligator River. It is worth spending some time here, especially when nearby fig trees are fruiting. Birdwatchers regularly see the Rainbow Pitta and Little Shrike Thrush and sometimes the Cicadabird.

The Bardedjilidji Sandstone walk passes through sandstone outcrops, open woodland and wetlands along the East Alligator River which is lined with paperbarks and patches of monsoon forest. These habitats may reveal Orange-footed Scrubfowl, Emerald Dove and Rainbow Pitta.

E. South Alligator

Mamukala has two bird hides on a walk around a wetland area. It is an excellent place to look for waterbirds, such as the Little Curlew. The waterside trees normally house Restless and Lemon-bellied Flycatchers and the Barking Owl roosts in the Freshwater Mangroves near the first hide. The walk goes through open woodland with good opportunities for seeing Long-tailed, Masked and Crimson Finches and Chestnut-breasted Mannikin.

The Arnhem Highway crosses the South Alligator River at an extensive floodplain. The mangroves near the boat ramp can produce Azure Kingfisher and the tall grass hides Zitting Cisticola, which is common and easily located by its call in the wet season. This area is also a haunt for the Yellow Chat.

The Gungarre walk travels around the Kakadu Holiday Village through monsoon forest to Anggardabal billabong and then back though open woodland. This walk — brilliant for birdwatching — is easy and is open all year. Start at the large fig tree on the highway to the east of the village. In the monsoon forest watch for Rose-crowned Fruit-Dove, Green-backed Gerygone, Little Shrike-thrush, Varied Triller and orioles. This is also another good billabong site.

It is worth stopping at any of the three branches of the West Alligator River and searching the dense vegetation for Azure and Little Kingfisher and possibly Nankeen Night Heron and Black Bittern. The Black-necked Stork, Brolga, and sometimes the Yellow Chat occur on the plains.

Other Areas

Other places are worth exploring mainly because they are less heavily visited. Red Lily Billabong, Alligator Billabong, Burdalba, Two Mile Hole and Four Mile Hole are good examples of other wetlands. Black Jungle on the Old Darwin Road features a good patch of monsoon forest. At Twin and Jim Jim Falls you can reach patches of monsoon forest near the escarpment. Both these sites are popular during the dry but you need a four-wheel drive vehicle to reach them.

4. LITCHFIELD

Background

Featuring four major spring-fed falls, Litchfield National Park is part of the Tabletop Range. The falls flow all year round and spill into clear pools surrounded by monsoon forest. Other habitats include escarpment and woodland. The park has a number of walking trails of varying standard and swimming is possible at the pools below the falls.

The Checklist contains 197 species for this area.

Access

Reach the park by a sealed road via Batchelor township or gravel road via Cox Peninsula Road. The formed gravel road provides dry season two-wheel drive access but may be corrugated and rough. The roads to the sites described are all sealed although the park contains some four-wheel drive roads. A day visit to Litchfield from Darwin is possible but a longer visit is worthwhile for birdwatching.

Accommodation

The nearby Batchelor township provides accommodation. The town will often have a congregation of birds attracted to the water supplies and flowering trees and is worth a visit, particularly in the late dry season.

Camping

The camping ground at Wangi Falls has full facilities, the two camping grounds at Florence Falls (one has only four-wheel drive access) and the one at Buley Rockhole have basic facilities. Caravans are only allowed at Wangi Falls but Batchelor has commercial caravan parks.

Litchfield

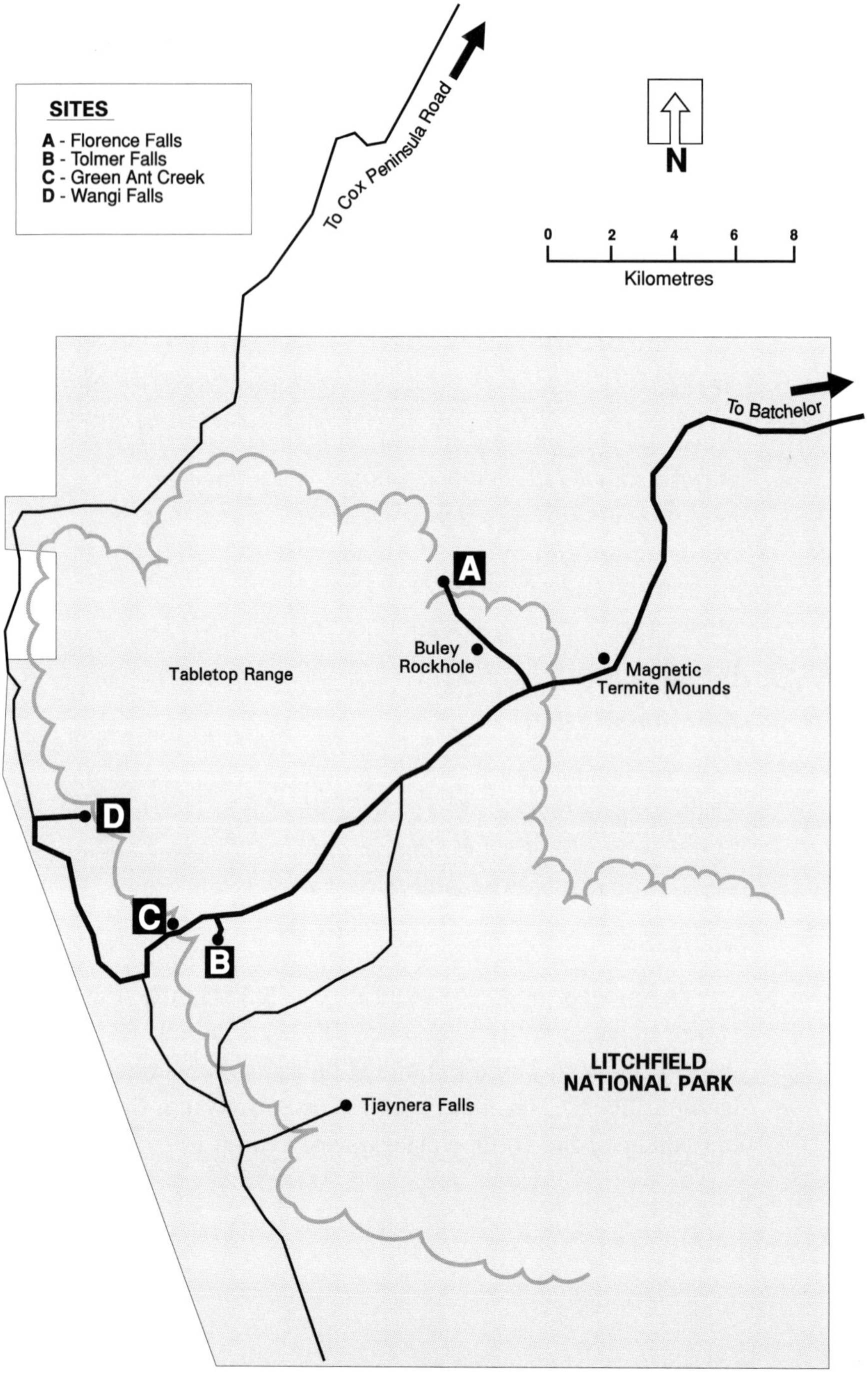

SITES
A - Florence Falls
B - Tolmer Falls
C - Green Ant Creek
D - Wangi Falls
To Cox Peninsula Road
N
0 2 4 6 8
Kilometres
To Batchelor
A
Buley Rockhole
Magnetic Termite Mounds
Tabletop Range
D
C
B
LITCHFIELD NATIONAL PARK
Tjaynera Falls

Special Birds

	Recommended Sites			
	A	B	C	D
Red Goshawk	•			
Partridge Pigeon	•			
Varied Lorikeet	•			
Northern Rosella	•			
Black-tailed Treecreeper	•	•		
Bar-breasted Honeyeater		•		

Sites

The sites are listed in order of entering the area from Batchelor, travelling in a clockwise direction and leaving via the Cox Peninsula Road.

A. Florence Falls

This is the best birdwatching site in Litchfield. It is a popular tourist destination and the swimming hole can become crowded from mid morning to mid afternoon, particularly in the dry season. A viewing platform close to the carpark looks out over the falls and monsoon forest, giving good opportunities to see Pied Imperial-Pigeon, Great Bowerbird, and possibly Rose-crowned Fruit-Dove.

The hill surrounds near the two-wheel drive campground are a good place to see Grey and Pied Butcherbirds. Great Bowerbird is common in the camping grounds. Black-tailed Treecreeper and Cicadabird are seen in this area and Partridge Pigeon is common on either recently burnt ground or patches of quartz outcrops at the base of the hills. The rocky hill slopes offer the best opportunity to see Varied Lorikeet, Northern Rosella, Leaden Flycatcher and others.

The Shady Creek Walk through margins of monsoon forest next to the stream gives good opportunities to see Green-backed Gerygone, Azure Kingfisher, Large-billed Gerygone, Little Shrike-thrush, Shining Flycatcher and Spangled Drongo. Watch for Little Kingfisher close to the water. White-throated Gerygone is common on the fringe of the monsoon forest.

A drive at night with a spotlight or car headlights will reveal Bush Stone-curlew, Southern Boobook, Spotted Nightjar and Australian Owlet-nightjar along the sides of roads and tracks. Dirt roads offer the best chances.

B. Tolmer Falls

A short walk from the carpark to a lookout offers spectacular views of the waterfall. An alternative loop walk follows the creek where grevilleas attract an abundance of honeyeaters and other nectar and pollen feeders. You may see Bar-breasted Honeyeater and in the wet season Azure Kingfishers are common. This walk gives access to a good variety of woodland birds including the Black-tailed Treecreeper.

One of the largest colonies of the rare Orange Horseshoe Bat inhabits the gorge and hence access to the gorge itself is prohibited.

C. Green Ant Creek

A walking trail starts at the Green Ant Creek carpark. Walk alongside well vegetated creeks surrounded by woodland, then climb uphill to the top of a waterfall fringed by paperbarks.

You will see a variety of woodland and monsoon forest species on this picturesque walk. Pied Imperial-Pigeon, Common Koel, Pheasant Coucal, Little Friarbird, Dusky Honeyeater and Shining Flycatcher are likely.

D. Wangi Falls

Good well-marked paths at Wangi Falls are worth a visit but may be crowded. These tracks pass through monsoon forest and woodland. Bar-shouldered Dove, Blue-winged Kookaburra, Silver-crowned Friarbird and Great Bowerbird are all common.

5. KATHERINE

Background

The Katherine area provides excellent opportunities for woodland birdwatching as well as providing monsoon forest and escarpment habitats. Nitmiluk National Park features the Katherine Gorge, which is carved from the Arnhem Land plateau, and Edith Falls.

The Checklist contains 220 species for this area.

Access

Katherine is located on the sealed Stuart Highway and is easily accessible by car or bus.

Accommodation

The town contains a range of accommodation from hotels and motels through to backpacker hostels.

Camping

Katherine Gorge and Edith Falls both have commercial caravan and camping grounds and swimming areas. The town also has commercial caravan parks.

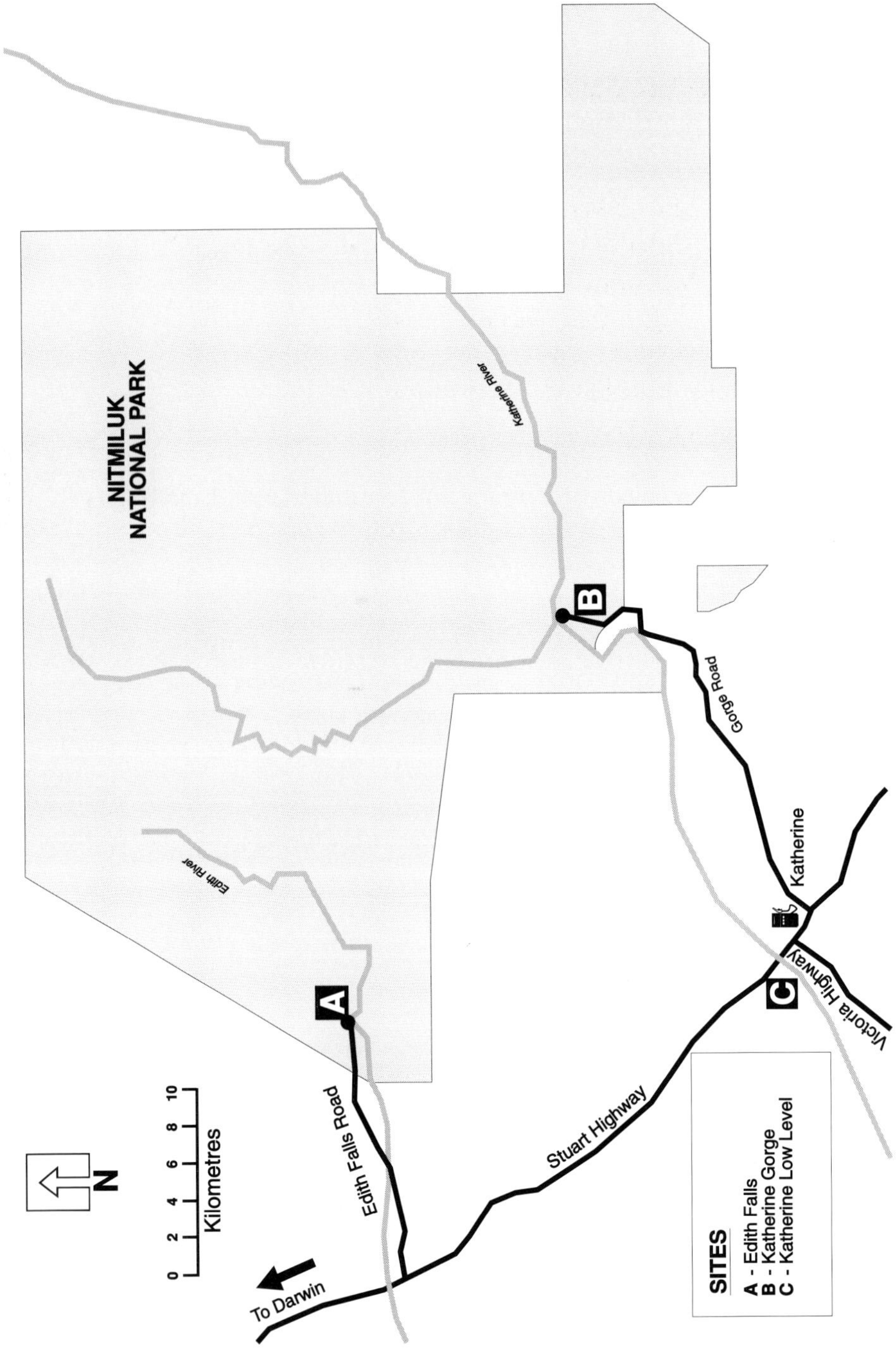
NITMILUK
NATIONAL PARK
Katherine River
Edith River
A
B
C
Gorge Road
Katherine
Victoria Highway
Stuart Highway
Edith Falls Road
To Darwin
0 2 4 6 8 10
Kilometres
N
SITES
A - Edith Falls
B - Katherine Gorge
C - Katherine Low Level

Special Birds

	Recommended Sites		
	A	B	C
Great-billed Heron		•	
Red Goshawk		•	
Hooded Parrot	•		
Black-tailed Treecreeper	•		
Bar-breasted Honeyeater	•		
White browed Robin	•		
Gouldian Finch	•		

Sites

A. Edith Falls

At Edith Falls, a large permanent waterhole with cascading waterfalls is fringed with monsoon forest, pandanus and rock walls. Hooded Parrot and Gouldian Finch inhabit salmon gums on the nearby hilly country.

During the dry season it is worth searching for Northern Rosella, Hooded Parrot, Black-tailed Treecreeper, Long-tailed Finch, Masked Finch and Gouldian Finch in areas just outside the park. You can see the finches and parrots coming to water early in the morning in the creeklines cutting through the steep hill country. Some of the small waterholes are near the roadside. The Northern Rosella and Hooded Parrot are noisy in the morning and should be easy to locate along roadsides. Pairs of Hooded Parrot will come in to drink during the heat of the day, but they are usually silent at this time. The Gouldian Finch is also silent and is best located by looking among the flocks of other finches, particularly the Long-tailed Finch, coming to water at early to mid-morning.

The Partridge Pigeon is often seen along the Edith Falls road, particularly near burnt ground, in the early morning and late afternoon. The Black-tailed Treecreeper is common in the area and you should be able to locate it by its call. Watch for the Black-chinned Honeyeater, Leaden Flycatcher and Great Bowerbird.

The main pool below Edith Falls, and the creek that flows from it are excellent places to see a great variety of honeyeaters. This place, as well as further upstream and downstream, is fringed with Fern-leaved Grevilleas which flower brilliant orange from May through to September and attract an incredible variety of species. Find a stand of these trees in flower (try near the main pool) and you will pick up many honeyeater species, including Bar-breasted, Rufous-throated and Banded. The creekline also houses

Nankeen Night Heron, Black Bittern, Grey Goshawk, Barking Owl, Azure Kingfisher, Blue-winged Kookaburra, gerygones and Crimson Finch and is a good place for flycatchers. Just on dark, Darter and cormorants roost on the exposed branches near the waterfall.

The main pool is also a good site for predator species like the Brown Goshawk, Peregrine Falcon and, occasionally, the Grey Goshawk. The Little Woodswallow can usually be seen soaring above the main pool.

A marked loop track leaving from the carpark climbs the escarpment. Watch for Sandstone Shrike-thrush on the rock face and Broad-billed Flycatcher in the vegetation fringing rock-pools. In the riverine vegetation look for Helmeted Friarbird, particularly during the dry season, and White-browed Robin.

The camping ground provides excellent opportunities to see typical woodland species such as the Bush Stone-curlew, Tawny Frogmouth, Australian Owlet-nightjar, Blue-winged Kookaburra and Great Bowerbird at close quarters.

B. Katherine Gorge

The Katherine River dissects the sandstone escarpment forming the spectacular Katherine Gorge which is the main attraction in the Nitmiluk National Park. Apart from the sheer beauty and exceptional cultural significance of the area, it also provides excellent birdwatching because it houses species in the transition area between the Top End and more arid regions. Boat trips are available at the Gorge.

The camping grounds are good places to observe some of the bolder woodland species, such as the Double-barred Finch, Crimson Finch and Northern Fantail. A profusion of bird activity is usually associated with the green lawns, food scraps and dripping taps. Watch for Northern Rosella. Walks in the area are well marked. Those that traverse the escarpment are worth the effort to see Sandstone Shrike-thrush on the slopes and, occasionally, Variegated (Lavender-flanked) Fairy-wren, usually in similar habitat. Early morning boat cruises give the chance of seeing the Great-billed Heron but look well ahead when trying to see this species as they usually slip away silently. You can also see the Black Bittern, Grey Goshawk and Azure Kingfisher. The river is lined with paperbarks and pandanus and has similar species to Edith River. Red Goshawk lives in the park.

C. *Katherine Low Level*

Popular for swimming, walking and canoeing, this park is close to town. Picnic facilities are available and the area is popular with locals. Watch for Grey Falcon in the area, particularly in the dry season. The Black Falcon occurs on the heavily grazed paddocks nearby and the Australian Hobby is common at sunset, flying low and fast over tree tops in pursuit of birds, but also feeding on dragonflies and other insects. This site attracts a lot of common birds, as well as species associated with drier regions such as the Crested Pigeon, Galah and Yellow-throated Miner. The Black-breasted Buzzard is often seen in surrounding country.

6. VICTORIA RIVER DISTRICT

Background

The Victoria River District extends along the Victoria Highway to the Western Australia border. It encompasses the Victoria River (the Territory's largest river) and its tributaries. The Stokes Range is a raised sandstone plateau similar to the escarpment of Arnhem Land. The vegetation of the area is predominantly open woodland with a grassy understorey.

Gregory National Park, a relatively new park, is split into two areas, the eastern section and the western section. Four-wheel drive is required to visit many sites in the park but excellent birdwatching is possible close to the Victoria Highway.

The Checklist contains 259 species for this area.

Access

Reach the area by the sealed Victoria Highway from either Katherine or Kununurra in Western Australia.

Accommodation

Victoria River Crossing and Timber Creek provide motel accommodation.

Camping

Victoria River Crossing and Timber Creek have commercial caravan parks and Bullita Homestead, Big Horse Creek and Limestone Gorge have basic camping grounds.

Victoria River District

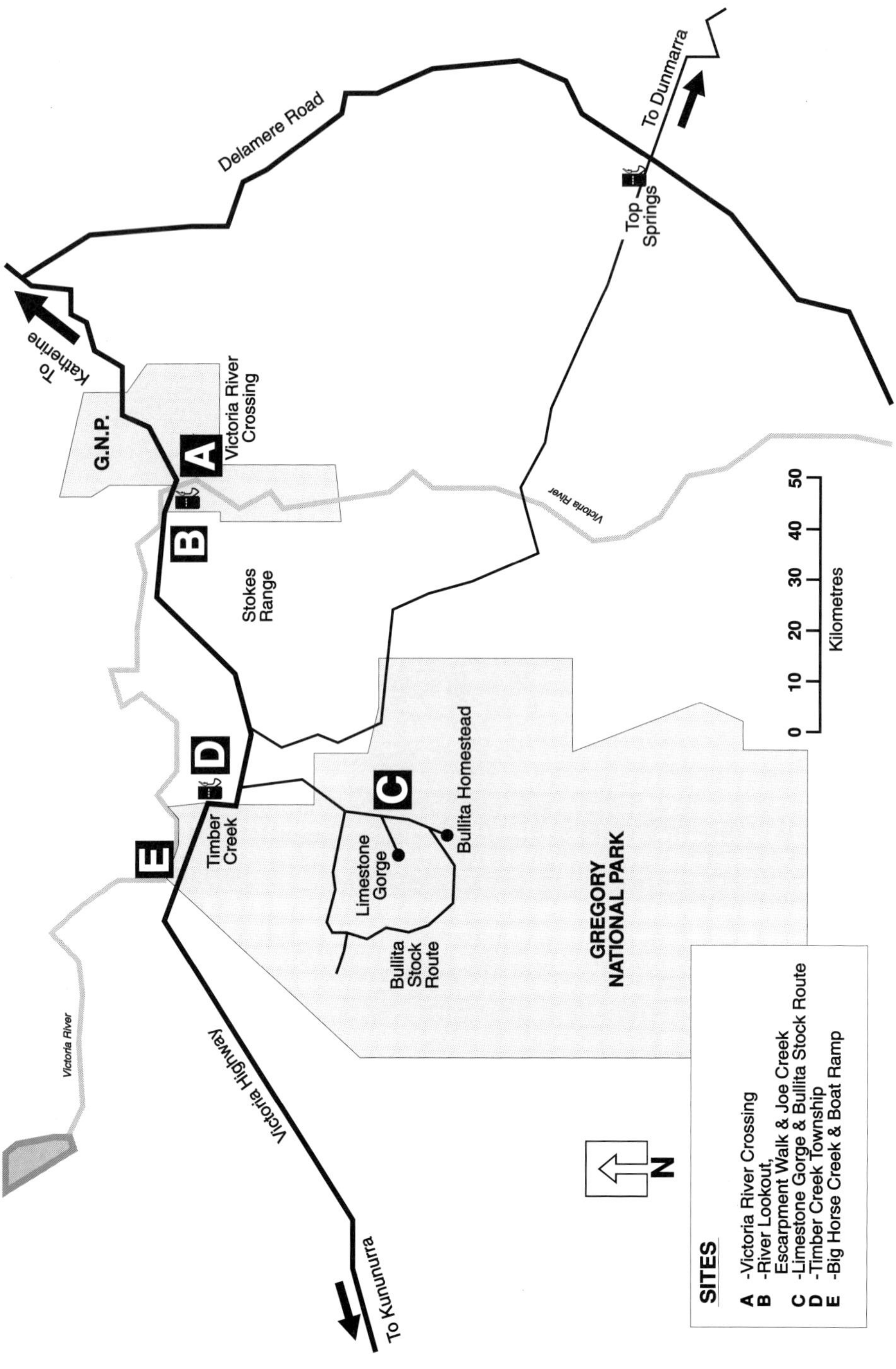

Special Birds

	Recommended Sites				
	A	B	C	D	E
King Quail				•	
Oriental Pratincole				•	
Purple-crowned Fairy-wren	•		•	•	•
Sandstone Shrike-thrush		•			
Star Finch	•			•	
Yellow-rumped Mannikin	•				
Pictorella Mannikin				•	
Gouldian Finch				•	

Sites

The sites described are listed roughly from east to west.

A. Victoria River Crossing

The area around the bridge at Victoria River Crossing gives easy access to the river and associated vegetation where you can see a good range of birds. This is an excellent place to look for the Purple-crowned Fairy-wren in the long grass, as access to the river and associated habitats is easy. Look for Star Finch and Yellow-rumped Mannikin feeding on seeding grasses with the common Crimson Finch and Chestnut-breasted Mannikin. The river is also worth looking at for waterbirds in the dry season when surface water becomes scarce. Raptors are present; search the thermals for Osprey and White-bellied Sea-Eagle.

B. River Lookout, Escarpment Walk and Joe Creek

These sites in the eastern section of Gregory National Park are all on the south side of the Victoria Highway. The River Lookout is the first walk you come to after leaving Victoria River Crossing. It climbs up the escarpment of the Stokes Range and provides beautiful views of the river. There is always a chance of seeing a White-quilled Rock-Pigeon, (which has similar habits to the Chestnut-quilled Rock Pigeon), the Sandstone Shrike-thrush and Little Woodswallow.

The Escarpment Lookout Walk is worth the effort as it provides access to the main Stokes Range with its escarpment and plateau. It is a good place for spotting White-quilled Rock-Pigeon and Sandstone Shrike-thrush.

On the Joe Creek walk, a further eight kilometres on, the loop trail climbs a slope and follows the base of the escarpment before descending to a valley floor and returning through open woodland. It is an interesting walk passing through patches of palms nestled at the base of the escarpment. Because the track follows the base of the escarpment, it may be hard to see the White-quilled Rock-Pigeon.

C. *Limestone Gorge and Bullita Stock Route*

These sites are in the western section of Gregory National Park. Limestone Gorge provides access to the escarpment and an area of lush, dense and varied riverine vegetation. A walk winds through limestone hills and passes a permanent swimming hole. Probably of most interest to the birdwatcher is the well-vegetated limestone creek and East Baines River. The Purple-crowned Fairy-wren occurs along these water courses and you can readily see the Little Shrike-thrush, Shining Flycatcher and other monsoon forest birds. The Partridge Pigeon inhabits roadsides and the expected finch species inhabit the grasslands.

Access to Limestone Gorge and Bullita Homestead is possible by two-wheel drive vehicle in dry conditions.

The Bullita Stock Route is definitely for those with a four-wheel drive vehicle and four-wheel driving experience. The route leaves the main park access road just north of the Bullita Homestead area. A registration book needs to be filled in and information leaflets are provided. After travelling in a clockwise direction around the route you re-encounter the main access road and sign out at another book. This route crosses limestone ridges as well as the East Baines River at a number of points. It goes through woodland, good areas of grassland habitat and riverine vegetation and passes some beautiful swimming holes.

D. Timber Creek Township

The town is surrounded by woodland and has many watering points, with the irrigated grasses and shade trees making an ideal habitat for finches. You can find all 11 species of Territory finches and mannikins in the Timber Creek area. The sought after species are Star Finch, Gouldian Finch, and Yellow-rumped and Pictorella Mannikins. It is worth walking along the main street looking in the grass on both sides of the road for the Star, Yellow-rumped and Gouldian species. Also check out Timber Creek itself behind the town, which is lined with pandanus and paperbarks and is home to many birds, including Barking Owl and Azure Kingfisher. A pair of Barking Owls lives along the creek near the western caravan park. Timber Creek is lined with good Purple-crowned Fairy-wren habitat so look in the more secluded areas.

As the dry season progresses, more finches come into town to drink from leaking taps and to feed on the irrigated lawns. Timber Creek township is a study site for the Gouldian Finch.

Besides the town itself, two other places worth searching for finches are the racecourse and the airstrip. The racecourse is 10 kilometres east of the town, just off the main road. Long-tailed, Masked and Star Finches and Pictorella and other mannikins are periodically common, and the Gouldian Finch is regularly seen here. Other species, such as the Grey-fronted and Black-chinned Honeyeaters are also abundant. The airstrip, west of town, is a finch haven, especially in the early dry when surface water is available and grasses are seeding. All the sought after finch species are found here, while Painted Finches live in the surrounding rocky country. When feeding in the long grass, finches can be quite hard to find, so keep an eye out for groups in flight and try to follow them.

The King Quail is a very localised bird which is found in damp soggy areas of grass. It has been seen in suitable habitat west of the township — listen for its calls.

The woodland around Timber Creek is a transition zone between semi-arid and tropical woodland species. You could find Oriental Pratincole here.

E. Big Horse Creek and Boat Ramp

To the west of Timber Creek, this popular camping spot is on the banks of the Victoria River. Once again, the predominant vegetation is woodland and all the associated birds inhabit the area. The banks of the Victoria River may also be a good place to look for the Purple-crowned Fairy-wren and for waterbirds in the late dry. You can see some wader species around the boat ramp area but watch out for crocodiles.

The site is excellent for night birds. The best method of spotting them is to drive along dirt roads during the first few hours after dusk. Southern Boobook and Spotted Nightjar are the most numerous on the roads. The Bush Stone-curlew is reasonably common and you may also see Barn Owls and the Australian Owlet-nightjar. Riverine vegetation is the best place to find the Barking Owl which comes out earlier than other owls and is very vocal around dusk (and dawn).

7. TOP END TO THE CENTRE

Background

This chapter will give you an idea of what is to be seen between the Top End and the Centre when travelling the 1200 kilometres from Katherine to Alice Springs. The chapter also describes a detour into the Barkly Tablelands.

The long distances require at least one overnight stop. Travelling by car at night is not recommended because of cattle and kangaroos wandering on the roads.

Because of its size and the large number of habitats covering both the Top End and the Centre, this Chapter is not included on the Checklist.

Access

All the main roads described are sealed although parts of the optional Barkly Tablelands detour are single lane bitumen roads that require particular care when passing or overtaking. The road into Connell's Lagoon is a good dirt road when dry.

Accommodation

Most of the roadhouses provide accommodation and Tennant Creek has hotel, motel and backpacker accommodation. Borroloola also has places to stay.

Camping

Caravan and camping facilities are available at the roadhouses along the way and in Tennant Creek and Borroloola. Camping is available at Twelve-Mile in Elsey National Park, at Longreach Waterhole and at the Devil's Marbles. Camping is not allowed at Connell's Lagoon or Caranbirini.

Top End to the Centre

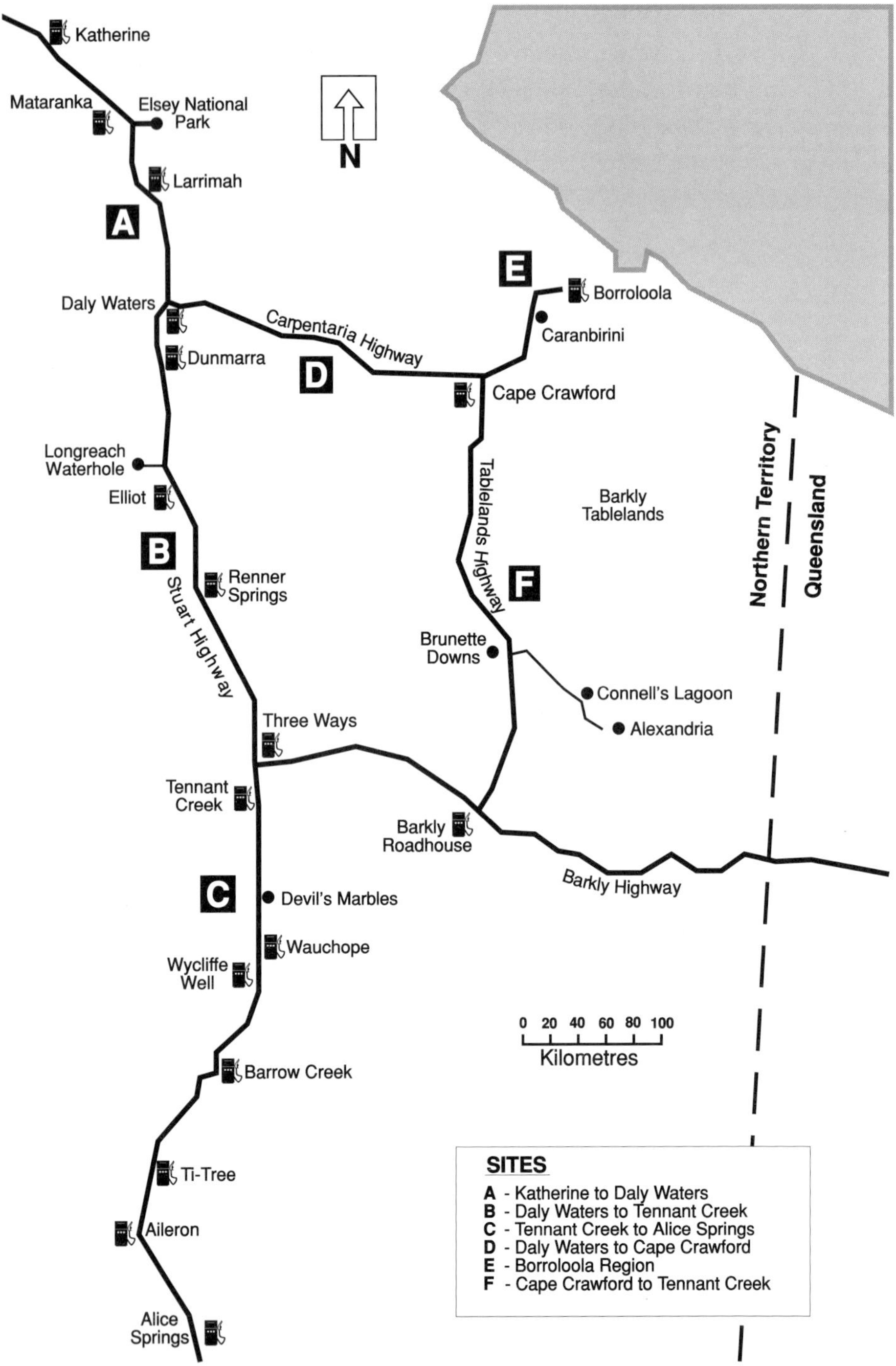

Special Birds

	Recommended Sites					
	A	B	C	D	E	F
Letter-winged Kite						•
Grey Falcon		•	•			
Oriental Plover		•			•	•
Oriental Pratincole		•			•	•
Flock Bronzewing		•				•
Purple-crowned Fairy-wren					•	
Carpentarian Grasswren					•	
Slaty-backed Thornbill			•			
Yellow Chat		•				•
Ground Cuckoo-shrike		•	•			•
Pictorella Mannikin						•

Sites

Sites are described as sections of the routes.

A. Katherine to Daly Waters

The habitat is primarily tropical woodland dominated by eucalypts, but as you move south the understorey becomes more varied and is increasingly dominated by shrubs and small trees rather than grasses. The birdlife, while still predominantly that of the Top End, begins to change in species composition and abundance. Some Top End species, such as the Varied Lorikeet, Red-winged Parrot, Yellow-tinted Honeyeater, Banded Honeyeater and Grey-crowned Babbler become more abundant. Others associated with the more arid areas start to appear, including the Crested Pigeon, Diamond Dove, Galah, Yellow-throated Miner, Singing Honeyeater and Zebra Finch.

Elsey National Park near Mataranka has hot thermal springs surrounded by monsoon forest and other areas dominated by woodland. The Rainbow Lorikeet, Blue-faced Honeyeater, Apostlebird and Great Bowerbird are common. The walking track from the Twelve-Mile camping area to Mataranka Falls follows the Roper River and will produce a range of species.

The beautifully coloured Black-chinned Honeyeater is patchily distributed in the interior, at times being reasonably common. It is never far from a watercourse.

B. Daly Waters to Tennant Creek

Birds associated with the Centre, such as the Cockatiel, Australian Ringneck, Red-backed Kingfisher, thornbills, Spiny-cheeked Honeyeater, Red-capped Robin and Hooded Robin become common. You may see some of the arid specialities such as the Painted Finch and Spinifexbird as well as the Grey Falcon along this stretch of road.

Longreach Waterhole, part of Lake Woods, is a waterfowl reserve and an important breeding ground for many waterbirds, such as the Brolga, Jabiru and Glossy Ibis. Also look for the Varied Lorikeet, Rufous-banded, Banded and Grey-fronted Honeyeaters, Jacky Winter, Restless Flycatcher and Ground Cuckoo-shrike. If you are lucky, you may see Flock Bronzewings and Yellow Chats. To reach the waterhole, turn west off the Stuart Highway immediately after the grid just north of Elliot and travel for about 12 kilometres along a dirt road.

Look at the flowering trees in Elliot for Singing, Grey-fronted, Brown, Rufous-throated and Banded Honeyeaters. You may also see the White-browed Woodswallow, a rare bird for the Northern Territory.

On the stretch from Elliot to Tennant Creek there is a chance to see the Spotted Harrier, Australian Bustard, Variegated and White-winged Fairy-wrens, Grey-headed Honeyeater, Richard's Pipit, Painted Finch, Spinifexbird and Rufous Songlark.

House Sparrows are present at Tennant Creek.

C. Tennant Creek to Alice Springs

The main habitat of this section is shrubland dominated by Mulga. Other areas have low rises covered with spinifex. The bird species composition is typical of the arid and semi-arid environments. Slaty-backed Thornbills live in the Mulga shrubland and can be confused with the very similar Chestnut-rumped Thornbill. The Devil's Marbles, about 100 kilometres south of Tennant Creek house the Painted Finch and sometimes, Spinifexbird and White-winged Fairy-wren. Along this section of road, you have a good chance of seeing Black Falcons and Black-breasted Buzzards.

Barkly Tablelands Route

First travel section A, then these sites and then section C.

D. Daly Waters to Cape Crawford

Along the Carpentaria Highway the woodland is varied and tends to have a dense shrub layer. It is easy to see a wide diversity of woodland species, such as, the Dollarbird, Black-tailed Treecreeper, White-throated Gerygone, Varied Sitella, Rufous Whistler, Grey Shrike-thrush, Apostlebird, lorikeets honeyeaters and flycatchers.

E. Borroloola Region

At Caranbirini Nature Park there is a good chance of seeing the Grey Goshawk, Purple-crowned Fairy-wren, Carpentarian Grasswren, White-browed Robin and, in the dry, the Gouldian Finch. Search in the pandanus and long grasses fringing the water for the fairy-wren and robin. The preferred habitat of the grasswren and Sandstone Shrike-thrush is the nearby rugged sandstone ridges covered in spinifex. The Star Finch, Chestnut-breasted Mannikin and Pictorella Mannikin are also present at Caranbirini and the Red Goshawk has been reported.

In Borroloola itself, the Purple-crowned Fairy-wren and White-browed Robin have been recorded near the Holiday Village and the school. The Mangrove Whistler lives in the nearby mangroves.

F. Cape Crawford to Tennant Creek

Travelling south along the Tablelands Highway to the Barkly Roadhouse and then west along the Barkly Highway to Three Ways just north of Tennant Creek, the route passes through some of the Barkly Tablelands, a place little visited by birdwatchers. This is the stronghold of the Letter-winged Kite and Flock Bronzewing and to a lesser extent the Yellow Chat. The Pictorella Mannikin also occurs in good numbers in this region and you may also see the Australian Brolga, Australian Pratincole, Ground Cuckoo-shrike and Singing Bushlark.

Connell's Lagoon is the only reserve in the Barkly Tablelands and one of the key reasons for its establishment was to conserve Mitchell grassland communities. Access is via a dirt road east of the highway which is impassible when wet. The habitat is mainly grassland with some shrubs and swamps and a number of bores. Notable birds of the park are the Letter-winged Kite, Australian Bustard, Red-chested Button-quail, Oriental Pratincole, Flock Bronzewing, Yellow Chat and Pictorella Mannikin.

The nocturnal Letter-winged Kite roosts in stands of trees in open woodland or grassland or along tree-lined watercourses. At night it disperses to the surrounding open woodland or grassland to hunt.

The Flock Bronzewing is a highly mobile species which, being a seed eater, needs to drink regularly. It approaches water in flocks in the morning and afternoon. Camouflaged while feeding on the ground, it is more often seen in flight.

8. ALICE SPRINGS

Background

Superb birdwatching opportunities exist close to Alice Springs. Located in the centre of the picturesque MacDonnell Ranges it hosts a wide variety of range and shrubland species, some of which are highly sought after. The Sewage Ponds are notable for rarities and vagrants.

The Checklist contains 191 species for this area.

Access

Alice Springs has a domestic airport and bus services from all states and Darwin. The sealed Stuart Highway runs from Adelaide to Alice Springs and then on to Darwin. The Ghan railway service operates from Adelaide.

Accommodation

The town has a wide range of accommodation from backpacker hostels through to luxury hotels.

Camping

A number of commercial caravan parks have camping facilities. No camping is allowed at the Telegraph Station.

Alice Springs

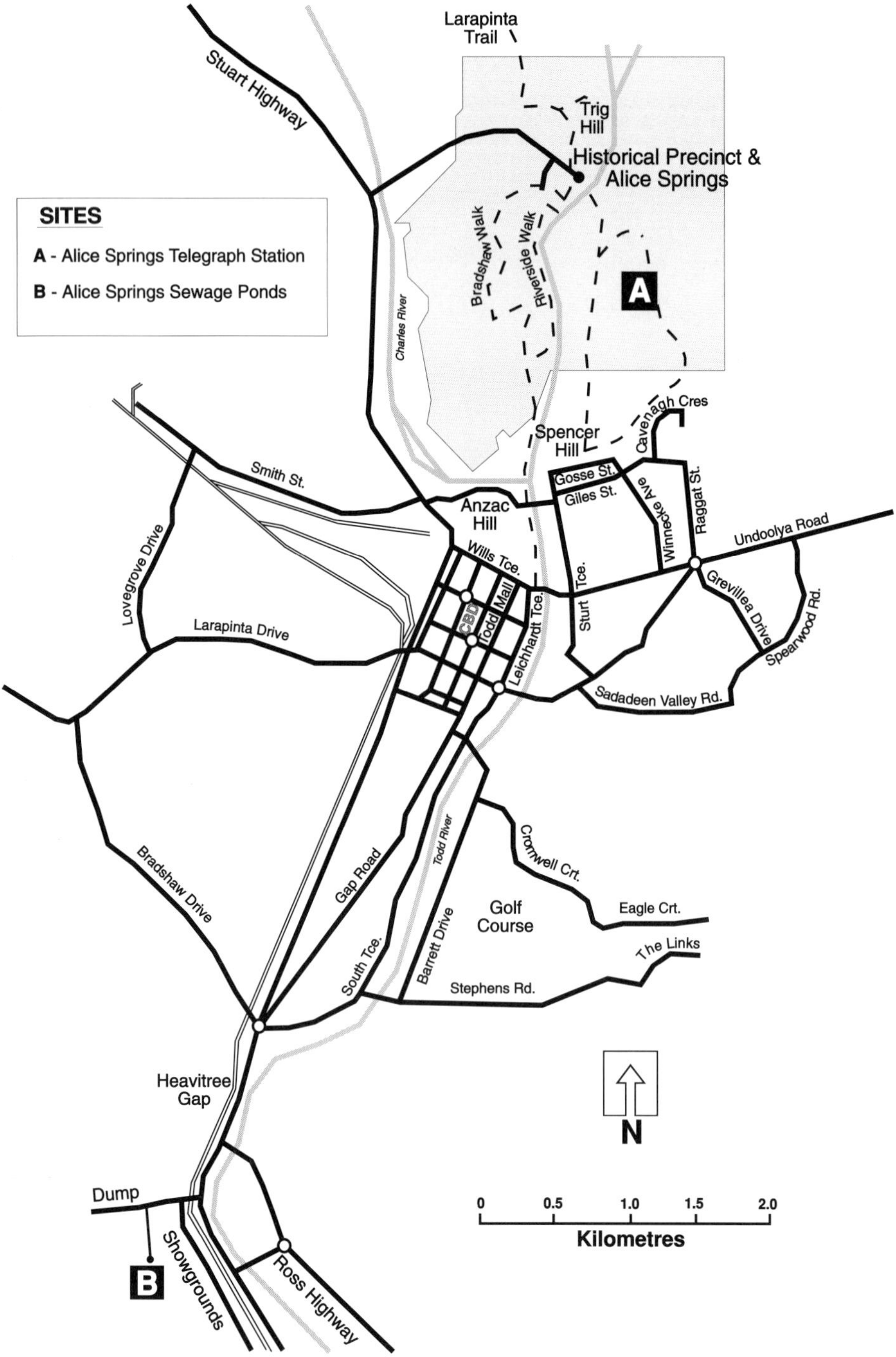

Special Birds

	Recommended Sites	
	A	B
Grey Honeyeater	•	
Black Honeyeater	•	•
Pied Honeyeater	•	•

Sites

A. Alice Springs Telegraph Station

The Telegraph Station at the Alice Springs waterhole is the site of the original European settlement in the area. If driving, head north on the Stuart Highway and take the signposted turn to the right. A number of walking tracks from town travel along either side of the Todd River and the Telegraph Station is the start of the long distance Larapinta Trail through the West MacDonnell ranges. Entry to the reserve and picnic areas is free but a fee is charged to enter the historic building area. The reserve is open from 8 am to 7 pm from May to September, and 8 am to 9 pm from October to April.

The Telegraph Station on the banks of the Todd River has well-watered lawns that attract a number of birds including Australian Ringneck, Grey-crowned Babbler and Zebra Finch. The River Red Gums on the river banks house Red-browed Pardalote and honeyeaters, and provide nesting hollows for parrots. Raptors are common here, including Black and Whistling Kites, goshawks, Wedge-tailed Eagle, Australian Hobby, Little Eagle, Nankeen Kestrel and occasionally Black-Breasted Buzzard and Black Falcon.

The walking trails through hilly country in the south-eastern section of the reserve provide good opportunities to see many shrubland species. Horsfield's Bronze-, Pallid, and Black-eared Cuckoos are conspicuous, vocal and common after good rains when other species are breeding. At this time Crimson Chats arrive in good numbers, while White-fronted, Black and Pied honeyeaters are prevalent in the flowering eremophilas. Diamond and Peaceful Doves are common. The rare Grey Honeyeater is said to be resident here. The spectacular Splendid Fairy-wren is seen mainly in shrubs on rocky hilly country where Western Gerygone, Inland, Chestnut-rumped and Yellow-rumped Thornbills and Red-capped Robin also occur. The Redthroat is resident all year round: an excellent mimic, it will be noticed first by its loud, beautiful mimicry which can sound as though many birds are within a single shrub. The Masked Woodswallow can appear in large numbers. The Crested Bellbird is conspicuous and the Australian Bustard, Spinifex Pigeon, Mulga Parrot, Fork-tailed Swift, Hooded Robin and Ground Cuckoo-shrike appear sometimes.

B. Alice Springs Sewage Ponds

The sewage ponds are open to the public for birdwatching although they are operational ponds, not a recreation area. To get there, head south out of town on the Stuart Highway, take the first turn to the right, south of the gap and then the second turn on the left after the railway line.

The ponds provide a large permanent water source and birdlife is abundant throughout the year. You can usually find the Australian Wood Duck, Pacific Black Duck, Grey Teal, Pink-eared Duck, grebes, cormorants, egrets, herons, Eurasian Coot, Black-winged Stilt, Red-necked Avocet, Black-fronted Dotterel, Masked Lapwing and terns. Also reported are the Plumed Whistling-Duck, Blue-billed Duck and Freckled Duck. Look for the Clamorous Reed-Warbler, Little Grassbird and crakes in the reeds around the ponds and Variegated and White-winged Fairy-wrens in the saltbush.

Summer migrants include the Marsh Sandpiper, Common Greenshank, Wood and Common Sandpipers, Red-necked Stint and Sharp-tailed Sandpiper. Anything could show up including Swinhoe's Snipe, godwits, Little Curlew, Long-toed Stint, Pectoral Sandpiper, Curlew Sandpiper, Ruff, Oriental Plover and Oriental Pratincole.

Black and Whistling Kites are common and the Australian Hobby and Peregrine Falcon sometimes come to harass the waterbirds; the Black Falcon is resident. The area around the ponds is worth searching for common shrubland species. Honeyeaters are common, and Black and Pied Honeyeaters have shown up in good years. A number of waterbirds breed in a swampy area to the south of the ponds and the Swamp Harrier regularly occurs here.

9. EAST MACDONNELL RANGES

Background

The East MacDonnells provide excellent access to range, riverine and shrubland habitats and the quartzite ranges provide a sensational backdrop to the area.

The Checklist contains 133 species for this area.

Access

Take the Ross Highway out of Alice Springs. The highway to the Arltunga turnoff is sealed and access to Ross River Homestead, Arltunga Historical Reserve and Trephina Gorge is on good dirt roads. N'Dhala Gorge and the Arltunga tourist route require a four-wheel drive vehicle. A high clearance vehicle is recommended for John Hayes Rockhole.

Accommodation

Ross River Homestead and the Arltunga Pub provide accommodation.

Camping

Basic camp sites are available at Trephina and N'Dhala gorges. The Ross River Homestead and the Arltunga Pub have commercial camping grounds.

East MacDonnell Ranges

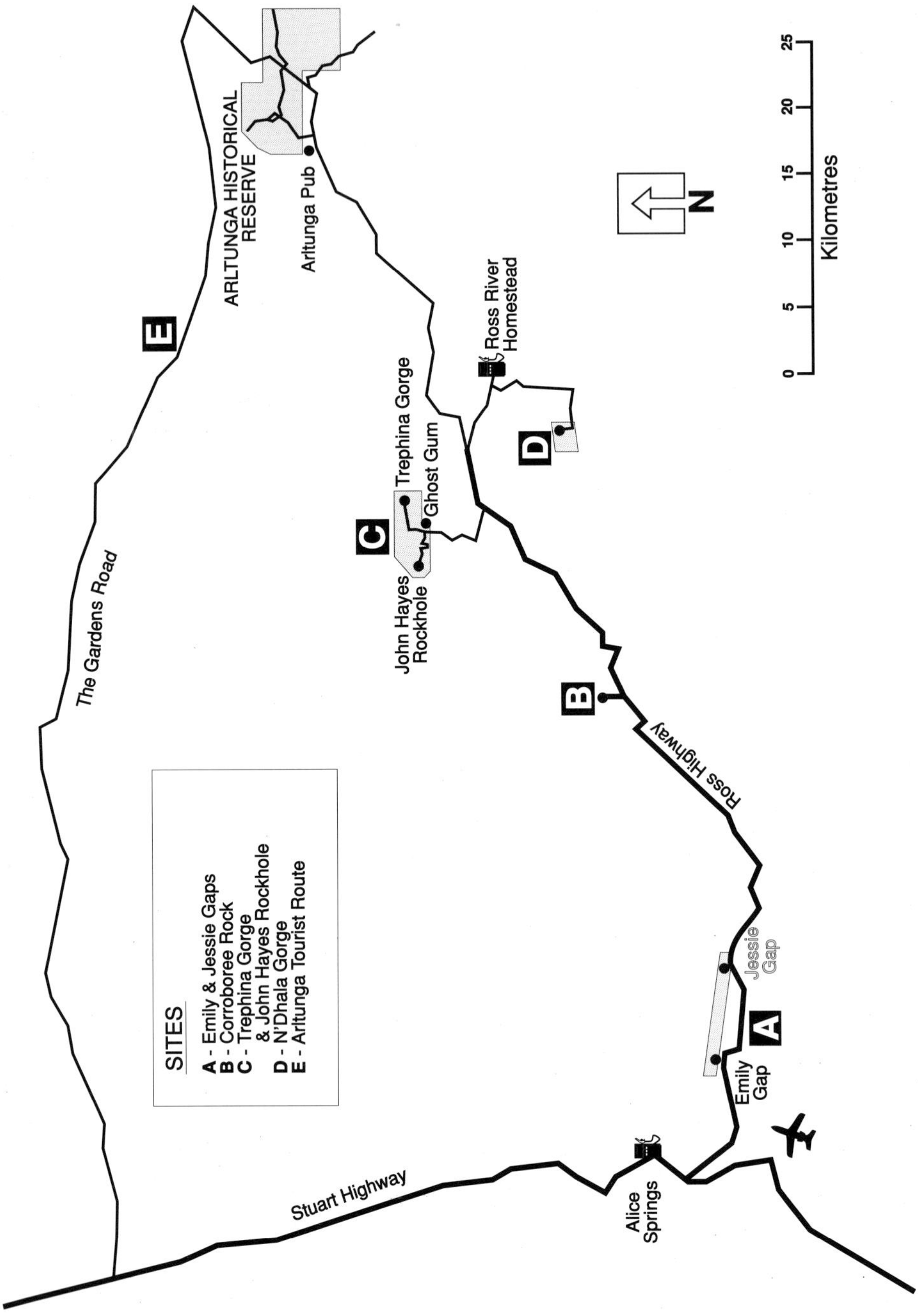

Special Birds

	Recommended Sites				
	A	B	C	D	E
Flock Bronzewing					•
Spinifex Pigeon	•		•	•	
Bourke's Parrot					•
Dusky Grasswren	•		•	•	
Banded Whiteface					•
Black Honeyeater			•		•
Pied Honeyeater	•		•		•
Ground Cuckoo-shrike					•
Painted Finch	•		•	•	
Spinifexbird		•	•		

Sites

The sites are listed from Alice Springs out along the Ross Highway into the East MacDonnells and returning to Alice Springs via the Arltunga tourist route.

A. Emily and Jessie Gaps

Reasonably close to Alice Springs, these two gaps are popular for picnics. You will find Red-backed Kingfisher, Red-browed Pardalote, honeyeaters, Mistletoebird and other riverine species. The Dusky Grasswren lives in the hills, and there is a chance of seeing the Black-chinned Honeyeater and Channel-billed Cuckoo.

B. Corroboree Rock

A short walk around the base of the rock gives access to spinifex, in parts with an overstorey of mallee, and a well vegetated creek-line. The species are mainly common ones such as the Grey-headed Honeyeater and Black-faced Woodswallow, however it is worth looking for Spinifexbird.

C. Trephina Gorge and John Hayes Rockhole

Sandy creekbeds, rugged range scenery and semi-permanent waterholes make this site a popular destination. Follow a number of marked walking tracks of varying standard.

At the John Hayes Rockhole, look for the Red-backed Kingfisher, the Dusky Grasswren, the Red-browed Pardalote and the Hooded Robin. While sitting near the waterhole you will see Spinifex Pigeon, Diamond and Peaceful Doves, Zebra Finch and possibly Painted Finch. In the late afternoon and early morning you may see Peregrine Falcon, Cockatiel and Black-chinned Honeyeater. The Chain of Ponds Walk is an easy track which travels through gorge and range habitat. A quick search in shrubs near the giant Ghostgum may yield Splendid and Variegated Fairy-wrens.

The Trephina Gorge campground provides good opportunities for exploring riverine, shrubland and range habitats.

Many walks begin at the carpark at Trephina Gorge. The Trephina Gorge walk ventures through the creek and back along the top of the gorge where you could add Brown Falcon, Common Bronzewing, Australian Ringneck, Little Woodswallow and Western Bowerbird to your site list.

For the more adventurous and better prepared the one-way four and a half hour Ridgetop Walk travels from Trephina Gorge to John Hayes Rockhole through range habitat consisting mainly of spinifex and mallee. This is ideal habitat for Spinifexbird.

D. N'Dhala Gorge

The dirt road leading to this gorge crosses the Ross River and passes through riverine and shrubland habitat before arriving at the gorge. A marked walk leads up the gorge where Galah, Little Corella, Rufous Whistler, Painted Finch and Mistletoebird are common. Mistletoe is common in the surrounding shrubland and, when flowering, attracts White-fronted and other honeyeaters.

E. Arltunga Tourist Route

This four-wheel drive route takes you through more than 100 kilometres of hilly country, shrubland, plains and riverine habitat. The Arltunga Historical Reserve contains ruins from gold mining days and is a good birdwatching area because of the varied shrubland. Along the Tourist Route watch for Ground Cuckoo-shrike on grassy flats where at times they are flushed from the side of the road. Little Button-quail can turn up in good numbers. Flock Bronzewing has been recorded here so carefully scan the plains and look ahead as you drive along. Inland Dotterel, Banded Lapwing and Australian Pratincole inhabit open stony plains and the Banded Whiteface flies around shrubs on these plains. Black Falcons are often seen surveying the open country from perches in the taller trees. Watch for Bourke's Parrot in shrubland. This is a place where Black and Pied Honeyeaters regularly turn up after good rains.

10. WEST MACDONNELL RANGES

Background

The rugged West MacDonnell Ranges extend westwards from Alice Springs for hundreds of kilometres. Much of the area is now contained in the West MacDonnell National Park. Comprising many permanent natural waterholes, spectacular gorges and towering cliffs, this area may well provide the birdwatching highlight of the Centre.

The long distance Larapinta Trail will eventually extend from the Alice Springs Telegraph Station through the West MacDonnell National Park, visiting Simpson's Gap, Standley Chasm, Ormiston Gorge and Mount Sonder. Several sections are already complete and information and brochures are available from the Parks and Wildlife Commission of the Northern Territory.

The Checklist contains 164 species for this area.

Access

Namatjira Drive is sealed to Glen Helen, as are access roads into Simpson's Gap, Standley Chasm, Ochre Pits, Ormiston Gorge and Glen Helen Gorge. The roads to Ellery Creek Big Hole and Redbank Gorge are unsealed but accessible by two-wheel drive vehicle unless wet.

Accommodation

Dormitory through to motel standard accommodation is available at Glen Helen Lodge next to Glen Helen Gorge.

Camping

Ormiston Gorge has a camping ground with full facilities; Ellery Creek and Redbank Gorge have basic camping grounds; Glen Helen Lodge has a commercial camping ground and camping is allowed at the Two-Mile area although no facilities are provided.

West MacDonnell Ranges

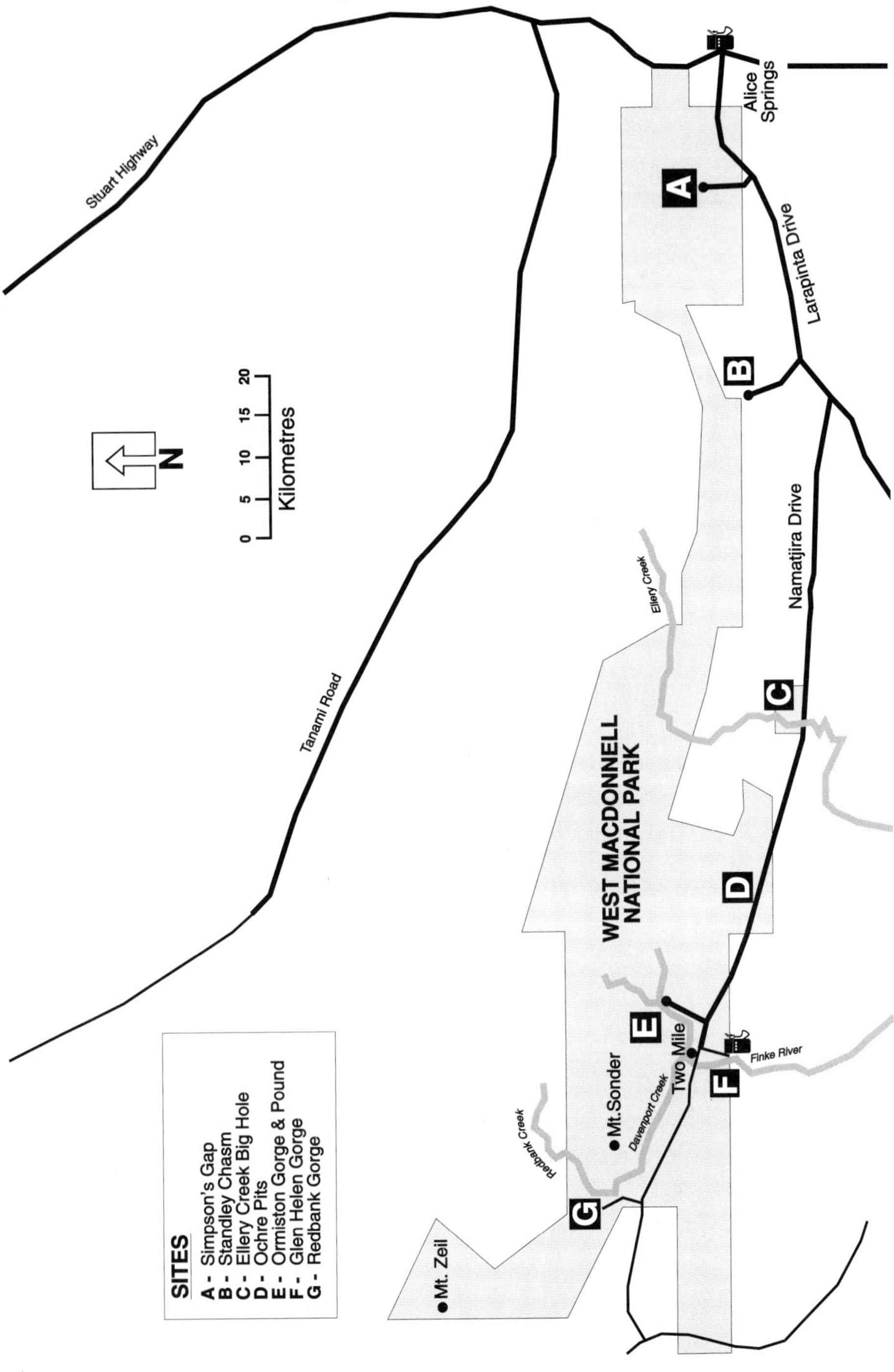

Special Birds

	Recommended Sites						
	A	B	C	D	E	F	G
Spinifex Pigeon		•	•		•	•	•
Bourke's Parrot	•		•				
Rufous-crowned Emu-wren			•	•	•		
Dusky Grasswren	•		•	•	•		•
Grey Honeyeater	•						
Black Honeyeater			•	•	•	•	
Pied Honeyeater			•	•	•		
Western Bowerbird	•	•	•	•	•	•	
Painted Finch	•	•	•		•	•	•
Spinifexbird			•	•	•		

Sites

The sites are listed from east to west along Larapinta and Namatjira Drives.

A. Simpson's Gap

It is worth having a look at Simpson's Gap (open from 8 am to 8 pm daily), although because it is so close to Alice Springs it tends to be more crowded than the more distant sites. The bike path from John Flynn's Grave to Simpson's Gap and back is a pleasant day's ride on which you will see numerous honeyeaters. Look out for Variegated and White-winged Fairy-wrens, the Grey-headed Honeyeater, woodswallows and, after rain, Pallid Cuckoo and Crimson Chat. Among several walks near Simpson's Gap, the Woodland and Cassia Hill walks are both excellent for shrubland species such as Redthroat, thornbills and Rufous Whistler.

The walk into Simpson's Gap follows a sandy creek with large River Red Gums. You will regularly see Dusky Grasswrens and occasionally, a Painted Finch. A pair of Peregrine Falcons lives at the gap itself and Zebra Finches are numerous; Bourke's Parrot occasionally drinks from the waterhole.

Section Two of the Larapinta Trail starts at Simpson's Gap and travels west. Crested Bellbird and thornbills are common and Grey Honeyeaters have been reported along this section of the trail.

B. Standley Chasm

An entrance fee applies at Standley Chasm, which is open from 8.30 am to 4.30 pm daily. At a shop near the carpark, some very tame Grey Shrike-thrush hop around tables and visitors' feet. Several species appear on the walk along the creek to the gorge including the Common Bronzewing, Little Woodswallow, Torresian Crow and Painted Finch.

C. Ellery Creek Big Hole

With its permanent water, large trees and surrounding spinifex-covered hills, the Ellery Creek Big Hole is excellent for birds — it is also a great place for a swim. Parrots, fairy-wrens and robins live in the nearby mallee shrubland. Cuckoos are seen in the area, and watch for Brown Goshawk and Black-chinned Honeyeater along the creekline. Budgerigar, Sacred Kingfisher, Rainbow Bee-eater and Little Woodswallow are often seen above the waterhole and Grey Falcons have been recorded along the creekline.

Rufous-crowned Emu-wren, Dusky Grasswren and Spinifex-bird are regularly found on the Dolomite Walk which starts near the shade shelter at the entrance to the waterhole. Bourke's Parrot has also been recorded in the area.

D. Ochre Pits

This is another worthwhile birdwatching site. The natural rock colours are spectacular and interpretive signs tell how Aboriginal people use the ochre. It is a recommended site for Rufous-crowned Emu-wren and Spinifexbird. The best place to look for them is on the Arrernte Walk that begins at the creekbed about 100 metres along the ochre cliffs. About three hours is required for the walk which finishes at Inarlunga Pass, where you can find the endemic MacDonnell Ranges Cycad. Thornbills, Varied Sitella, Western Bowerbird and Painted Finch are among the birds you will encounter on this walk.

Red-backed Kingfisher, Splendid and Variegated Fairy-wrens, Red-browed Pardalote, thornbills, honeyeaters, Western Bowerbird, Zebra Finch and Mistletoebird all appear in the acacia shrubland surrounding the carpark.

E. Ormiston Gorge and Pound

A spectacular feature of the MacDonnell Ranges, Ormiston Gorge is one of the area's best birdwatching sites.
A permanent waterhole provides a welcome dip after a hot

day, along with an opportunity to watch a variety of birdlife including ducks, herons, grebes and cormorants. Black-fronted Dotterels are always present at the water's edge.

This is a good base for exploring the West MacDonnells with excellent camping facilities, a variety of walking tracks and many birds.

In the Mulga shrubland surrounding the camping area, Spinifex Pigeon, parrots, honeyeaters, Rufous Whistler and Western Bowerbird are common. This is probably the easiest place in Australia to see Spinifex Pigeons which are tame in the carpark and camping grounds, and entertaining as they come in to feed, drink and chase each other.

As the surrounding environment dries up, large numbers of Painted Finch converge around the remaining water sources. Feeding in the surrounding slopes in the spinifex, they call loudly as they come in to the small pools for water.

A walk along the road parallel to the riverbed is likely to be rewarding in the early morning. You will often find Sacred Kingfisher, both species of martins, and raptors, such as the Brown Goshawk, Collared Sparrowhawk and Peregrine Falcon. Major Mitchell's Cockatoo has lived in the creek from time to time. This is also a good place to look for the Black-footed Rock Wallaby which is easily seen by a quiet observer. They live in the rocks and hills beside the road.

The Ormiston Pound Walk travels through spinifex-clad hills and crosses vegetated creeklines before entering the Ormiston Creek. You can find Rufous-crowned Emu-wren, Dusky Grasswren and Spinifexbird along the track. The occasional Black-chinned Honeyeater may appear in the Bloodwoods, while the Grey-headed Honeyeater is common in the Bloodwoods and Ghost Gums.

Larapinta Trail Stage 10 begins at the Visitor Centre and travels through spinifex, riverine and shrubland habitats to

Glen Helen, providing excellent opportunities for viewing Dusky Moorhen, Red-tailed Black Cockatoo and Rufous-crowned Emu-wren.

F. Glen Helen Gorge

The Finke River cuts through the ranges near Glen Helen Homestead forming the picturesque Glen Helen Gorge where a permanent waterhole provides an excellent habitat for waterbirds. Access to the southern side of the gorge is difficult but is possible by swimming through the waterhole or climbing around the rocks at its edge. The waterbirds prefer this end as they are less disturbed by visitors. Occasionally, species that are rare in the Centre, such as Osprey, Buff-banded Rail and Baillon's Crake, will appear. The Eurasian Coot, Dusky Moorhen, Purple Swamphen, Hardhead, Darter, cormorants and other waterbirds are usually present; while the Pink-eared Duck, Sharp-tailed, Common and Wood Sandpipers turn up from time to time.

Shallow waterholes which are worth visiting occur upstream from the gorge at the Two-Mile area. A good variety of birds can be found including ducks, egrets, herons, spoonbills, dotterels, waders, Red-tailed Black-Cockatoo, White-backed Swallow, Clamorous Reed-warbler, Rufous Songlark and, occasionally, an Emu. Access is difficult and particular care is required in two-wheel drive.

G. Redbank Gorge

A dirt road, requiring caution in two-wheel drive, leads to Redbank Gorge. This is quite a remote part of the West MacDonnells and camping here is ideal for those seeking isolation. The woodland bush camping area is some distance from the gorge itself but offers the best birdwatching. Located on a floodplain, the area has good riverine vegetation surrounded by Mulga shrubland where you can see birds such as parrots, Redthroat, Western

Gerygone, thornbills, honeyeaters and both species of babblers. Red-capped Robin and Weebill are common.

Redbank is a long narrow gorge with deep icy waters. You enter it after a 20 minute walk up the sandy creekbed from the carpark. A good way to experience its eerie stillness is by floating through on an air mattress. On the way in you may see Peregrine Falcon, Red-browed Pardalote, honeyeaters and Painted Finch.

Flowering mistletoe in the mallee shrubland surrounding the carpark provides habitat for Grey-headed and White-fronted Honeyeaters and Little Woodswallow.

11 WATARRKA

Background

As well as some interesting birds, Watarrka National Park contains the spectacular Kings Canyon. The surrounding sandy country is dominated by Desert Oak with a spinifex understorey. After rain the country looks brilliant with a profusion of annual flowers.

The Checklist contains 121 species for this area.

Access

To reach Watarrka, follow the Ernest Giles Road (the eastern half is not sealed) off the Stuart Highway or from Uluru, take the sealed Luritja Road. From the West MacDonnells, you have two dirt road options: the two-wheel drive Mereenie Loop (Site E, a permit is required) and the four-wheel drive Boggy Hole road through the Finke Gorge National Park (Site F).

Accommodation

The Kings Canyon Resort contains accommodation ranging from backpacker rooms through to luxury motel rooms.

Camping

The Kings Canyon Resort and Kings Creek Station have commercial camping grounds; Palm Valley has a camping ground with full facilities and camping is allowed at Boggy Hole (no facilities).

Watarrka

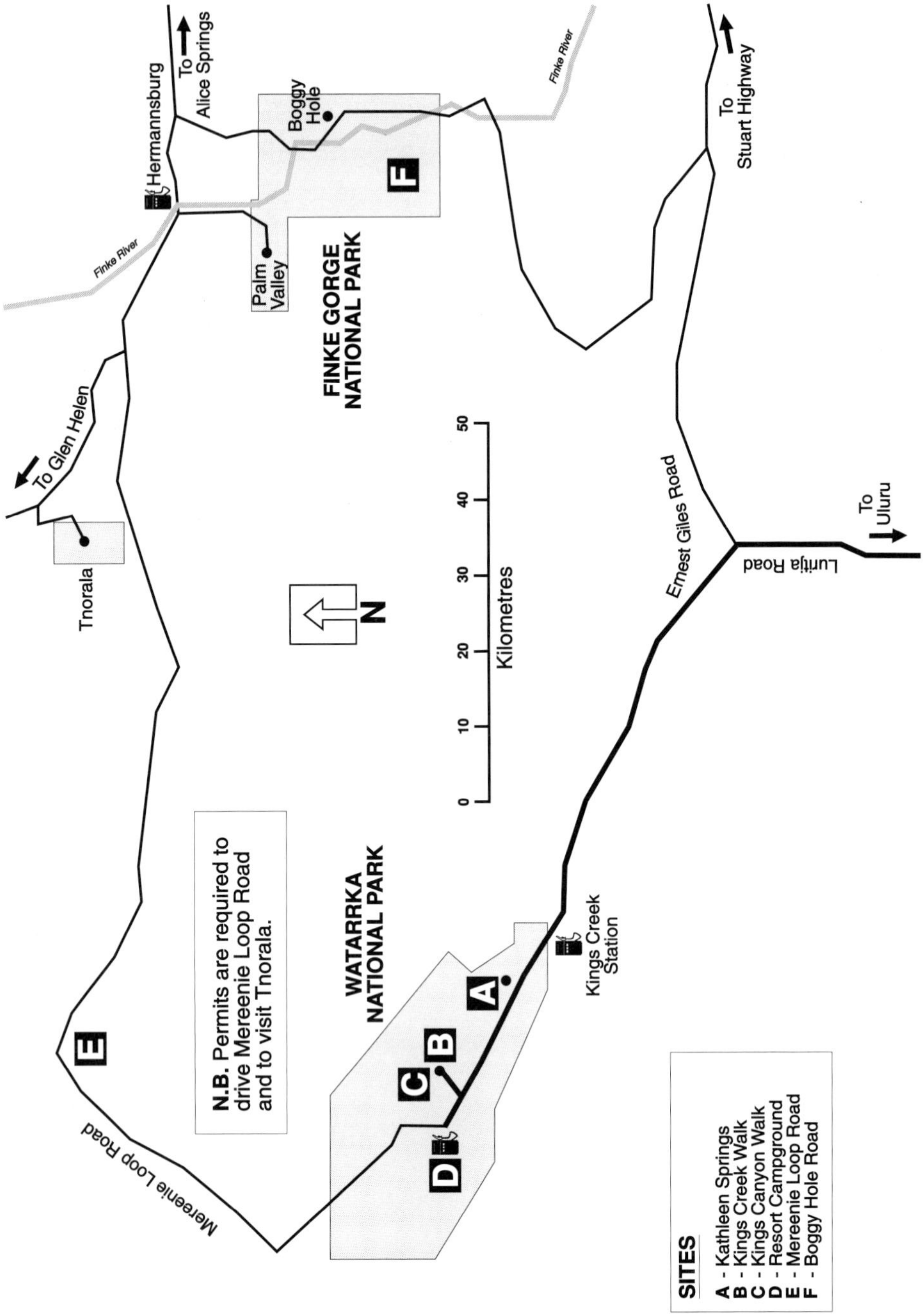

To Alice Springs
Hermannsburg
Boggy Hole
Finke River
Finke River
F
Palm Valley
FINKE GORGE NATIONAL PARK
To Glen Helen
Tnorala
To Stuart Highway
N
0
10
20
30
40
50
Kilometres
Ernest Giles Road
To Uluru
Luritja Road
N.B. Permits are required to drive Mereenie Loop Road and to visit Tnorala.
WATARRKA NATIONAL PARK
Kings Creek Station
A
B
C
D
E
Mereenie Loop Road
SITES
A - Kathleen Springs
B - Kings Creek Walk
C - Kings Canyon Walk
D - Resort Campground
E - Mereenie Loop Road
F - Boggy Hole Road

Special Birds

	Recommended Sites					
	A	B	C	D	E	F
Spinifex Pigeon	•	•	•			•
Dusky Grasswren	•	•	•			•
Western Bowerbird	•	•	•		•	•
Painted Finch	•	•	•			•

Sites

Sites are listed in order of their birdwatching potential, beginning with the best one.

A. Kathleen Springs

This short walk, which travels through shrubland to a spring-fed waterhole, offers good wheelchair access. The carpark and picnic area have extensive stands of dense Mulga with a range of birds typical of this habitat, including thornbills, White-fronted Honeyeater, Red-capped Robin, Hooded Robin and Crested Bellbird. The walk continues into a pound with a River Red Gum-lined creek. The riverine vegetation houses Red-browed Pardalote; and Black-eared Cuckoo in the understorey. The walk is good for honeyeaters with Grey-headed Honeyeater and White-plumed Honeyeater being common. The surrounding rocky country provides opportunities for Spinifex Pigeon, Dusky Grasswren and Painted Finch. The large number of raptors includes goshawks, Australian Hobby and at times Little Eagle. The trail ends at a small lagoon which should be a good place to watch birds coming to drink in the morning.

B. Kings Creek Walk

Starting at the main Kings Canyon carpark, the walk follows the creekline to the cliff base of the canyon. It provides a good opportunity to see the birds of rocky country, such as the Spinifex Pigeon, Dusky Grasswren and Painted Finch. The riverine vegetation houses common species and the walk takes you closer to the rocky country than the one at Kathleen Springs. The Painted Finch is difficult to see against the rocks but can be located by its incessant calling. It often stays around water during the day. The Western Bowerbird lives in the riverine vegetation of the gorge.

C. Kings Canyon Walk

Also starting at the main carpark, this walk requires about four hours and the steep climb at the beginning (good for Dusky Grasswren) can be quite strenuous. The trail, which is entirely through range habitat, usually contains a large number of walkers. Watch for Spinifex Pigeon and Painted Finch. Good eye-level views of the Little Woodswallow are possible. A detour through the Garden of Eden leads to a swimming hole which may have a few birds if it is not too crowded. The creekline leading to the swimming hole attracts birds such as the Grey-headed and White-plumed Honeyeaters and the Western Bowerbird.

D. Resort Campground

The watered lawns and planted trees at the campground attract large numbers of birds and provide a good place to see the common species of the area. Seven species of parrots and cockatoos occur nearby, including the Red-tailed Black-Cockatoo and Major Mitchell's Cockatoo. Watch for bird-eating raptors such as goshawk species, Little Eagle, Australian Hobby, Black-breasted Buzzard and Peregrine Falcon.

E. Mereenie Loop Road

Providing access to the West MacDonnells, the Mereenie Loop Road passes through Aboriginal land and travellers need to obtain a permit. These (including an informative brochure) are available at the Kings Canyon Resort, Glen Helen Homestead and a number of outlets in Alice Springs. The road passes through hilly areas, spinifex country and varied open shrubland. The route is well signposted and the permit contains additional information. Access to the sacred site of Tnorala (Gosse Bluff) is also covered by the permit. Tnorala has high-quality spinifex habitat and is a prime site for spinifex species including Rufous-crowned Emu-wren and Spinifexbird. The Western Bowerbird lives near the carpark and Red-backed Kingfisher, robins, and Black-faced and Little Woodswallows will be found in the nearby shrubland.

F. Boggy Hole Road

Definitely a four-wheel drive option, this track leaves the Ernest Giles Highway and travels north to join the Finke River in the Finke Gorge National Park. The track follows the river bed to Boggy Hole and eventually meets the Hermannsburg road from Alice Springs. The last part of the road is on Aboriginal land and while a permit is not required you should not leave the road. If you wish to visit Palm Valley, which is also in the Finke Gorge National Park, travel back to Hermannsburg, cross the Finke River and take the four-wheel drive road on the left. Along the way, many shrubland species are present: watch for Red-tailed Black-Cockatoo, Major Mitchell's Cockatoo, Cockatiel, Red-browed Pardalote and honeyeaters. Boggy Hole is a permanent waterhole and an excellent waterbird habitat. The diversity of waterbirds includes Australian Pelican, cormorants, coots, egrets and ducks. These are not included on the Watarrka checklist — refer to typical species found in the West MacDonnells.

12. ULURU

Background

This park is home to the famous Uluru (Ayers Rock) and Kata Tjuta (the Olgas). These features are surrounded by open desert vegetation which houses a wide range of common species. Although it contains some very remote country, the park has been relatively well researched and the species list is extensive. You probably won't see some of the birds on the Checklist such as the rare Princess and Scarlet-chested Parrots which turn up here very occasionally, their main range being in inaccessible country to the west.

The Checklist contains 169 species for this area.

Access

To reach Uluru, travel along the sealed Lasseter Highway. If you don't have your own vehicle, numerous bus tours are available or take one of the regular flights from Alice Springs and other states. An entry fee is paid as you enter the park or you can buy a Territorian Park Use ticket which allows unlimited access to Uluru-Kata Tjuta and Kakadu National Parks.

Accommodation

A wide range of accommodation is available at the Ayers Rock Resort. Make sure you book in advance.

Camping

The Ayers Rock Resort contains a commercial camping ground.

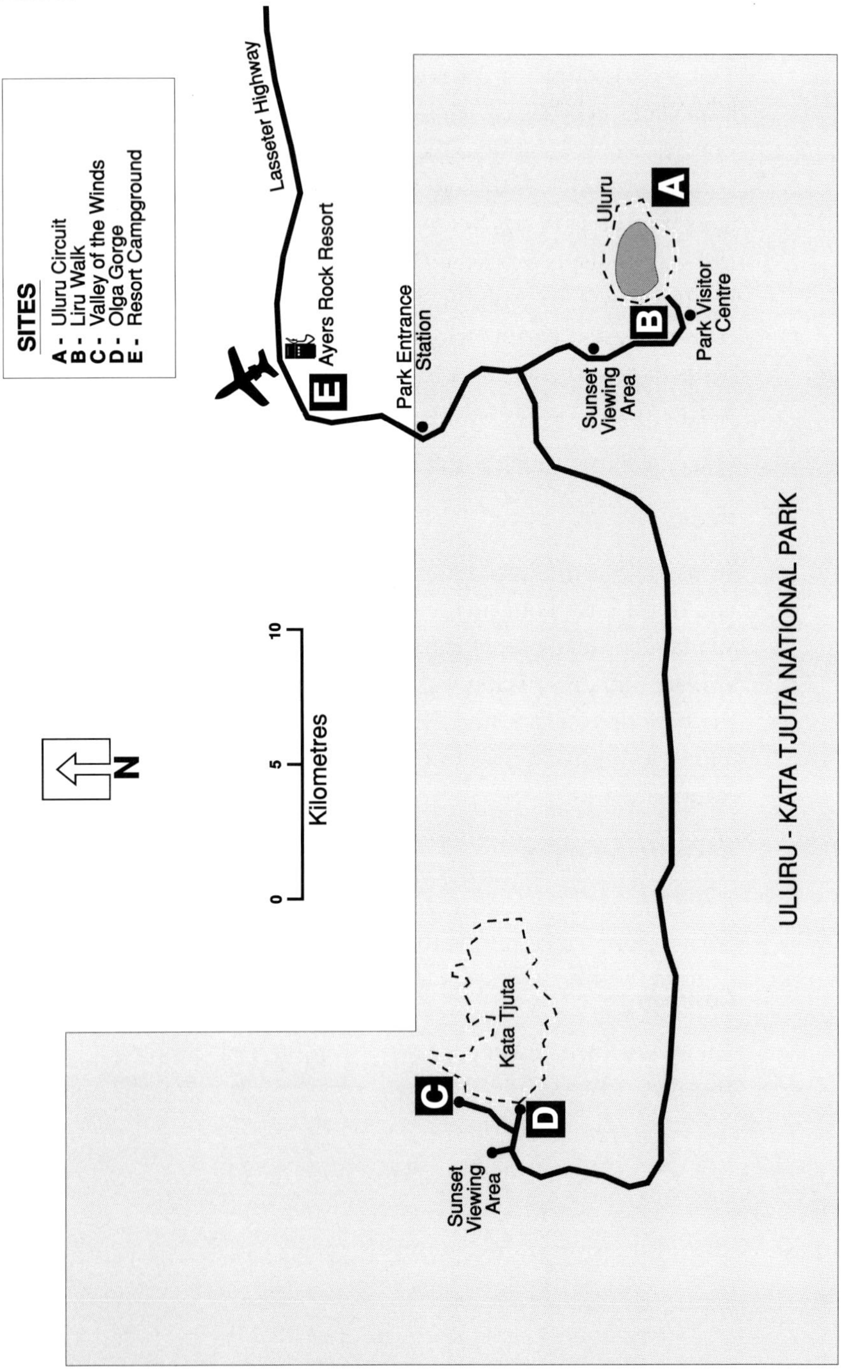
SITES
A - Uluru Circuit
B - Liru Walk
C - Valley of the Winds
D - Olga Gorge
E - Resort Campground
N
Kilometres
0
5
10
Lasseter Highway
E
Ayers Rock Resort
Park Entrance Station
Sunset Viewing Area
B
Park Visitor Centre
Uluru
A
Kata Tjuta
C
D
Sunset Viewing Area
ULURU - KATA TJUTA NATIONAL PARK

Special Birds

	Recommended Sites				
	A	B	C	D	E
Grey Falcon			•		
Oriental Plover	•				
Princess Parrot	•				
Grey Honeyeater	•				
Chiming Wedgebill			•	•	
Western Bowerbird	•	•	•	•	•

Sites

The sites described provide access to shrubland habitat and the rocky areas around the bases of Uluru and Kata Tjuta.

A. Uluru Circuit

This walk meanders around the base of Uluru. Travelling clockwise around the rock, the western and northern areas are predominantly Mulga scrubland and yield many birds such as wrens and thornbills. Hooded Robins perch in the taller scrub, often on dead branches offering a vantage point and you may also see Grey Honeyeaters. In good seasons quail-thrushes occur here.

After rain the grassier areas around the southern end of the rock are alive with Cockatiel, Mulga Parrot, Budgerigar and Little Button-quail. After winter rains the many breeding fairy-wrens and Crimson Chats in this area attract cuckoos.

The waterholes are a focus for many species during dry periods. Brown Goshawk and Collared Sparrowhawk will often hunt at these sites and Little Woodswallow and Mistletoebird are common throughout the area. Australian Hobby, Peregrine Falcon and Nankeen Kestrel nest and roost in the rock cavities and hunt in the surrounding country. Around the rock, Major Mitchell's Cockatoo is common in the Desert Oaks of the sand dune country and you may flush Oriental Plover from the side of the road in dry flat areas.

B. Liru Walk

Travelling from the Park Visitor Centre to the base of the climb through shrubland, expect to see common species for the habitat including thornbills, robins, honeyeaters and Western Bowerbird.

C. Valley of the Winds

This scenic walk travels around the domes of Kata Tjuta. Surprisingly, Grey Falcons have been regularly recorded along this walk while Peregrine Falcons are often seen perched in cavities in the rocks. Western Bowerbirds live in vegetation thickets and Little Woodswallows are commonly seen circling round the domes. In good seasons, you can find the Spinifex Pigeon and Painted Finch at the rock base. Dusky Grasswrens are rare here but you often see Grey and Pied Butcherbirds and Grey-headed and White-plumed Honeyeaters along the Bloodwood creeklines. Red-browed and Striated Pardalotes occur along the better vegetated creeklines, along with Black and Pied Honeyeaters in good seasons.

D. Olga Gorge

This shorter walk, which travels between Kata Tjuta's domes, is more vegetated than the Valley of the Winds walk. A reasonable number of birds inhabit the area, predominantly common shrubland species. The Western Bowerbird is resident while the Chiming Wedgebill is difficult to observe but betrayed by its chiming call.

E. Resort Campground

The lawns and water supply at Ayers Rock Resort attract numerous seed-eating birds which gather in large numbers, particularly Crested Pigeon, Galah and Zebra Finch, while honeyeaters, especially the Yellow-throated Miner and Spiny-cheeked and Singing Honeyeaters live in the planted eucalypts. In the evening these small birds attract the Brown Goshawk, Collared Sparrowhawk and Australian Hobby in pursuit of prey. The Little Crow is the common corvid here.

A walk to a sunset viewing vantage point behind the pool reveals fairy-wrens, chats and Rufous Songlarks in the sand dune habitat.

13. SOUTH EAST DESERT LOOP

Background

All the roads in this area are unsealed and at times you will need four-wheel drive. Visitors must be well prepared as there is little traffic and long distances between service stops. Maps, additional to the one found in this book, are essential. Most of the country is under pastoral lease so you should not wander away from the roadsides.

In good conditions birdwatching can be very rewarding with an abundance of birds but it is highly variable and greatly depends on local rain. Watch as you drive along and stop where birds flush from the side of the road. Crimson Chat is a good indicator species as it arrives in good seasons and is often common after rain.

The Checklist contains 159 species for this area.

Access

Follow the Alice Springs-Finke-Andado map and notes published by Westprint. The map is available from the Department of Lands, Planning and Environment in Alice Springs. Mount Dare is in Witjira National Park in South Australia.

Accommodation

Accommodation is available at Mount Dare. Some accommodation is available at the Old Andado homestead but phone ahead as it is not always attended.

Camping

Commercial camping facilities are available at Mount Dare and the Old Andado Homestead and a camping ground with basic facilities is located at Chambers Pillar.

South East Desert Loop

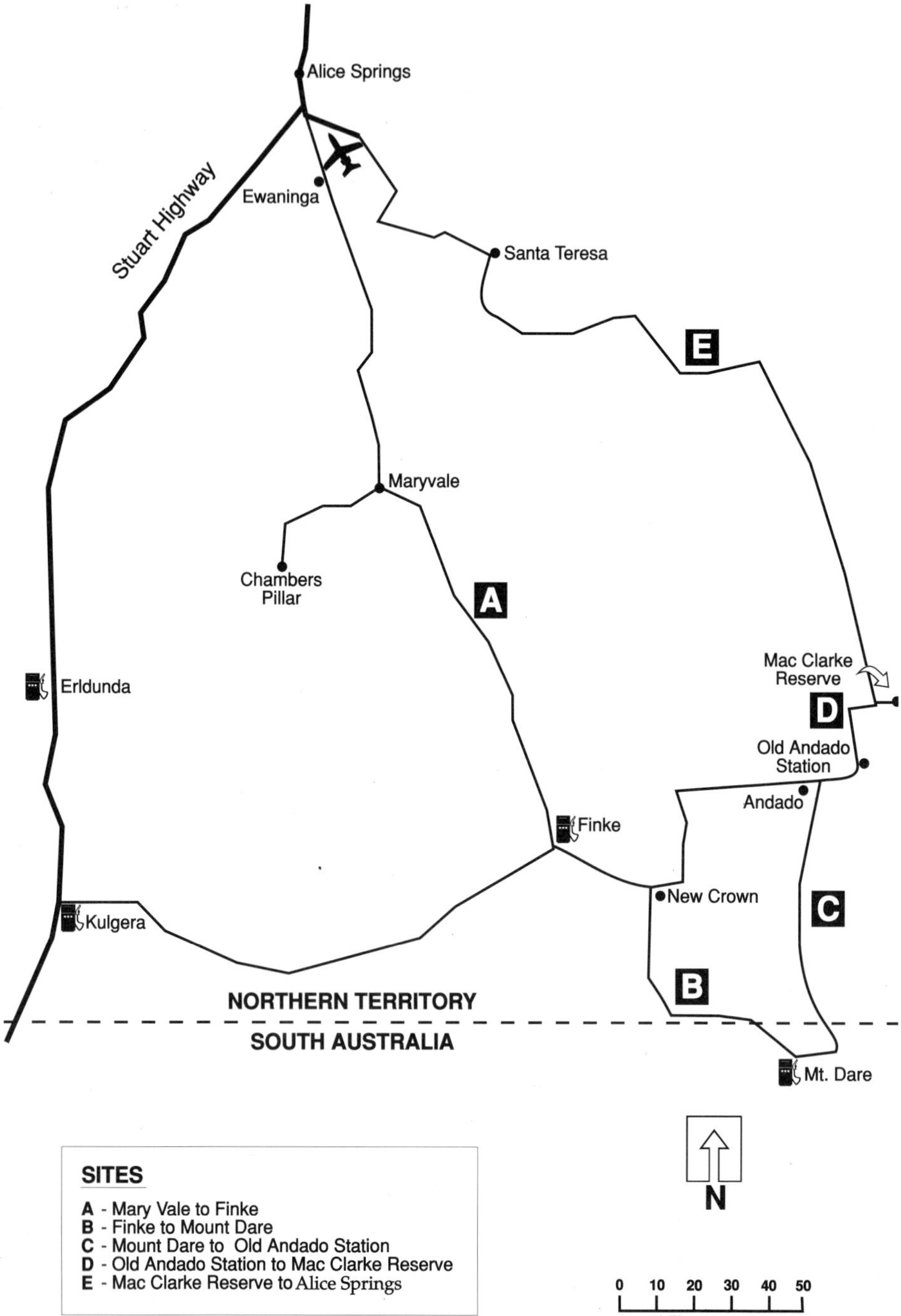

Special Birds

	Recommended Sites				
	A	B	C	D	E
Grey Falcon				•	•
Flock Bronzewing		•	•		
Bourke's Parrot	•	•			
Eyrean Grasswren			•	•	
Banded Whiteface			•	•	•
Gibberbird		•	•		•
Chiming Wedgebill				•	•
Ground Cuckoo-shrike	•		•	•	•

Sites

The sites are described in an anti-clockwise direction following the Alice Springs-Finke-Andado map (except that it goes straight from Finke to Old Andado and we detour via Mount Dare). Each section of the route is called a site.

A. Maryvale to Finke

The road from Alice Springs to Maryvale leaves from the road to the Alice Springs airport (follow the sign to Ewaninga Rock Carvings), is well formed and easy to follow. From Maryvale you can detour to Chambers Pillar (about 40 kilometres from Maryvale). The Maryvale to Finke section is a rough track along the old railway, passing through sand dunes with Desert Oak communities. You usually see Little Eagle, Black Falcon, Red-tailed Black-Cockatoo and Major Mitchell's Cockatoo. Just before Finke the road crosses the wide dry sandy bed of the Finke River and four-wheel drive is required.

Bourke's Parrot is common in acacia shrubland but search carefully as it is easily overlooked. A rarity for the Northern Territory, the Striated Grasswren, has been recorded here in spinifex with a sparse eucalypt overstorey. Cinnamon Quail-thrush is hard to find but reasonably common so listen carefully for its high-pitched call which is easily missed. Brown Falcon, Mulga Parrot, Red-backed Kingfisher and Crested Bellbird are widespread.

An easier alternative to this section of the route is to drive south to Kulgera on the Stuart Highway then along the formed dirt road to Finke.

B. Finke to Mount Dare

Drive south-east out of Finke towards New Crown Station where you turn south towards South Australia. The sandy country is heavily grazed but as the road passes into gibber country the bird watching opportunities improve. The road travels east towards Mount Dare roughly parallel to and just north of the Northern Territory–South Australia border.

Recently rediscovered in the Territory, the Thick-billed Grasswren can be found in the good chenopod vegetation of these gibber plains. It is unobtrusive and difficult to see as it scampers from bush to bush, well ahead of the observer.

If you move about quietly throughout the habitat and remain still and silent for a time, you will most likely encounter the Cinnamon Quail-thrush. The Rufous Fieldwren, another rarity for the Territory, also lives in the area. The Inland Dotterel is common in good seasons; during the day it relies on its camouflage but when closely approached it will walk away from an observer; at night it is active and will flush readily from the sides of roads. Watch out for Flock Bronzewing and Gibberbird. The plains provide good habitat for the Australian Bustard. The Orange Chat is a nomadic bird of chenopod shrublands which can appear in large numbers in good seasons and otherwise may be absent.

Bourke's Parrot lives in the Red Mulga creeks encountered before Mount Dare. Look for it in the early morning when it feeds on nearby plains, and in the evening when it returns to roost in the creeklines.

Mount Dare Homestead provides good watering points for birds, with doves, cockatoos, parrots and the abundant Zebra Finch. Hence raptors abound; Brown Goshawk, Australian Hobby, Black Falcon and Peregrine Falcon are constantly harassing their prey.

C. Mount Dare to Old Andado Station

The first part of this section passes through the Finke River floodout, a riverine habitat which is dominated by Coolibah trees with an understorey of shrubs or dense grasses. Rufous Songlark can be very numerous and cuckoos often use the habitat. Emu, quail, Spotted Harrier, Major Mitchell's Cockatoo, Australian Ringneck, Mulga Parrot, robins and Masked Woodswallow are all common. Rabbits have degraded the sand dunes close to the floodout so it is worth travelling a little further for good sand dune habitat.

After the floodout country the road passes through good sand dune country covered with spinifex and Sandhill Cane-grass. The cane-grass is easily recognisable as a wiry mass of stems that grow into large round clumps. Eyrean Grasswrens live in clumps of about one metre in height and in good seasons you often encounter these birds. The White-winged Fairy-wren and Rufous-crowned Emu-wren, which also inhabit the sand dunes, could be confused with the grasswren. Banded Whiteface and Cinnamon Quail-thrush inhabit interdunal swales. While driving along, watch for groups of Ground Cuckoo-shrike flushing from roadsides. Black Falcons often harass Galahs around waterholes and bores in the afternoon.

The road continues north and traverses an open gibber plain where Gibberbird is often flushed from the side of the road. Richard's Pipit, which also occurs in this habitat, may be confused with the Gibberbird.

Next to Old Andado Homestead is a large swampy area predominantly covered with Lignum, which only holds water after good rains. At these times a wide variety of birds can be observed including Pink-eared Duck, Hardhead, herons, egrets, Spotted Harrier, Swamp Harrier, Brolga, Australian Spotted Crake, Black-tailed Native Hen and, in summer, various waders. The nearby sand dunes house Eyrean Grasswren, and White-winged Fairy-wren is common.

D. Old Andado Station to Mac Clarke Reserve

This section passes predominantly through dunes covered with shrubs or Sandhill Cane-grass which are worth another look for Eyrean Grasswren. Listen for its call, it is very shy and elusive. All the associated sand dune species can be seen here. On well grassed flats, you can flush Little Button-quail.

The Mac Clarke Reserve is a small reserve protecting the rare Waddy Wood trees which exist on an otherwise bare plain. You may see Grey Falcons near the reserve while Inland Dotterel and Australian Pratincole can often been seen on the flat, almost featureless plains. Letter-winged Kite has been known to roost within the reserve and Yellow Chat has been recorded at bores in the area.

E. Mac Clarke Reserve to Alice Springs

After crossing more treeless country and passing some ranges, this road travels through shrubland vegetation dominated by acacias such as Mulga and gidgee. The habitat generally contains many birds, especially after good rains when trees and shrubs are flowering. At these times, White-fronted, Black and Pied Honeyeaters may be present in large numbers and you may see Chiming Wedgebill and Ground Cuckoo-shrike. Australian Bustard, Little Button-quail, Red-tailed Black-Cockatoo, Mulga Parrot, honeyeaters, Crimson Chat, White-winged Triller and woodswallows are common.

SPECIAL BIRDS

This section discusses birds considered to be special in the Northern Territory, including those which are generally easier to find in the Territory than elsewhere in Australia. The areas indicated are where we recommend that you look for the birds.
Special birds are also marked on the Checklist.

KING QUAIL

Darwin

Victoria River District

The King Quail is generally active at dawn and dusk. Its preferred habitat is dense, damp or swampy grassland in woodland or on floodplains. Listen for its persistent calling in the evening and early morning. When disturbed it prefers to run, taking flight only when hard pressed. When flushed it appears small and dark (especially the male), flies weakly with a heavy tail and plummets into ground cover.

GARGANEY

Darwin

Small Parks of the Top End

A rare but regular summer migrant, the Garganey can be found singly or in flocks in wetlands in the Top End. It takes flight easily and is difficult to approach. The bird is small (similar in size to Pink-eared Duck) and can be easily overlooked among other ducks. It appears most often at the Leanyer Sewage Treatment Ponds and Fogg Dam but it could turn up on any body of fresh water.

GREAT-BILLED HERON

Darwin

Kakadu

Katherine

The Great-billed Heron is mainly a bird of mangroves and riverine vegetation of the Top End. It is most easily observed feeding on exposed mudflats or flying along waterways in the early morning. When flushed in riverine vegetation it often flies up to perch on a large limb in the canopy. It is generally shy and easily disturbed by noise but in some areas it has become used to boats and can be easily approached.

BLACK BITTERN

Small Parks of the Top End

Kakadu

The Black Bittern is found primarily in pandanus and paperbark trees associated with fresh water. It tends to stay in the vegetation even when approached closely but will fly when pressed. Although it can become almost invisible in a large stand of pandanus, it is active during the day and is easier to see than other bitterns. Approaching its habitat by boat increases the chance of seeing it.

LETTER-WINGED KITE

Small Parks of the Top End

Top End to the Centre

The Letter-winged Kite's stronghold is the Barkly Tablelands although breeding populations have been recorded in the Adelaide and Mary River floodplains and recently in the vicinity of Fogg Dam. It could appear in any part of the Territory. The bird is nocturnal, becoming active on dusk and is difficult to find during the day because it roosts in leafy trees. Compared with the Black-shouldered Kite, its flight is more buoyant, its facial features appear owl-like and its underwing markings are quite different.

RED GOSHAWK

Kakadu

Litchfield

Katherine

The Red Goshawk is rarely seen and with approximately 350 pairs believed to exist in Australia, it is considered vulnerable. It often roosts along creeks and rivers, preferably those with a good margin of monsoon forest, and hunts in neighbouring woodland. Its favoured method of hunting is from a concealed perch but it will also soar above the trees, sometimes to a reasonable height, and generally in the warmer parts of the day.

GREY FALCON

Top End to the Centre

Uluru

South East Desert Loop

The Grey Falcon is a very rare nomad of inland Australia. It roosts and nests in riverine habitat and hunts on nearby open areas. Birdwatchers should carefully identify any possible sightings. The light phase Brown Falcon can be confused with the Grey Falcon but the former has distinct facial markings, brown legs and bill, rufous thighs and in flight, glides on upswept wings. The Grey Falcon may also be confused with the Brown Goshawk which has broader, more rounded wings. The Soaring Nankeen Kestrel can easily be misidentified but it has a slender build and a distinctive tail band.

WHITE-BROWED CRAKE

Small Parks of the Top End

The White-browed Crake's behaviour is typical of crakes. It keeps mostly to reeds and other aquatic vegetation but may feed in the open singly, in pairs, or in groups of up to 20 or more. It favours billabongs with reeds and lilies and at times uses the monsoon forest edges.

CHESTNUT RAIL

Darwin

Its loud calling and a series of strange grunting wheezing notes, often in duet, betrays the presence of the Chestnut Rail. This bird inhabits mangroves where it moves about on mud or walks out onto exposed mudflats searching for prey such as crabs. It is difficult to observe but low tide in the early morning gives the best chance. Be very quiet as it is secretive and easily scared by noise. It is possibly more common than reported in its habitat.

RED-BACKED BUTTON-QUAIL

Darwin

Kakadu

Three species of button-quail occur in the Top End, all in the same areas. They are difficult to flush and can be hard to identify. The Red-backed Button-quail is probably the most common. This is a small button-quail, much smaller than the Brown Quail. In the Top End it reportedly favours clay and black soil grassland with scattered shrubs and pandanus but it also occurs in sorghum beneath scattered eucalypts, and in spinifex along the Arnhem Land escarpment.

RED-CHESTED BUTTON-QUAIL

Darwin

Small Parks of the Top End

In the field the Red-chested Button-Quail appears similar in size to the Red-backed. Although the Red-chested prefers wetter areas with denser vegetation, the two are difficult to separate. The Red-chested is reported to be less common than the Red-backed.

CHESTNUT-BACKED BUTTON-QUAIL

Kakadu

This species is noticeably larger than the other two button-quails, being similar in size to the Brown Quail. Its prefers woodland habitat, with a grassy or spinifex understorey, on stony hillsides and sandstone ridges.

LITTLE CURLEW

Darwin

Small Parks of the Top End

Kakadu

Large numbers of Little Curlew enter Australia through Darwin Harbour and its surrounds during October and November and it can often be seen in the city at this time. It prefers dry short grassland and pastures and gathers on the floodplains in the Fogg Dam and South Alligator regions until after the first rains when it disperses mainly throughout northern Australia. Smaller numbers depart through Darwin in March and April.

BROAD-BILLED SANDPIPER

Darwin

The Broad-billed Sandpiper is a scarce summer migrant to coastal lagoons and beaches in the Top End. It prefers a muddy footing and mud banks of tidal creeks are a favoured feeding ground. It is uncommon and can be difficult to identify.

BEACH STONE-CURLEW

Darwin

Normally found singly or in pairs on exposed beaches and mudflats, the Beach Stone-curlew is reasonably common near Darwin. You often notice it first by its call and it is then easily observed. Usually wary, it takes flight readily and flies some distance. It is active at night and is often heard calling then. The bird is considered vulnerable nationally.

ORIENTAL PLOVER

Top End to the Centre

Uluru

The Oriental Plover is a summer migrant which arrives before the wet and leaves the Top End within a matter of days after the first rains, dispersing into the arid and semi-arid areas where it remains on dry open plains and shrub-land for the summer. It is a wary bird and not often seen.

ORIENTAL PRATINCOLE

Victoria River District

From the Top End to the Centre

Easily distinguished from the Australian Pratincole, the nomadic Oriental Pratincole can be seen in enormous flocks (usually in October and November) often associated with locust plagues, or prior to significant storms. At other times it is found in small numbers in open habitats associated with water, particularly in the Barkly region.

FLOCK BRONZEWING

Top End to the Centre

South East Desert Loop

The stronghold of the Flock Bronzewing is the Mitchell grasslands of the Barkly Tablelands. In the past it was seen in enormous flocks of many thousands but now it appears in flocks of hundreds. It is usually first seen flushing from the ground in the distance, or flying overhead. Its flight resembles that of the Rock Dove (feral pigeon) except that it has broader wings and does not generally glide. The colouring of the male is only visible with good views. It can be observed coming in to drink at dusk. It is a nomadic bird but possibly has a southerly movement in summer.

SPINIFEX PIGEON

East MacDonnell Ranges

West MacDonnell Ranges

Watarrka

The central race of the Spinifex Pigeon inhabits rocky hills and ranges covered in spinifex. Its beautiful rusty colours are a perfect camouflage against the sandstone and quartzite ranges. Being a seed eater it requires water regularly during the day, so the best way to find it is to wait near a rock pool. Occasionally it becomes tame in picnic areas and is well represented in the Territory's parks.

PARTRIDGE PIGEON

Kakadu

Litchfield

An uncommon and patchily distributed inhabitant of Top End woodland, the Partridge Pigeon is found on bare and burnt ground, especially in the early morning. It is often seen on the sides of roads and is easily approached. It usually occurs in pairs or groups of three to seven.

CHESTNUT-QUILLED ROCK-PIGEON

Kakadu

One of the Territory endemics, the Chestnut-quilled Rock-Pigeon feeds on seeds, roots and vegetation in the Arnhem Land escarpment country. It is well camouflaged and often the first indication of its presence is the loud "clapping" of its wings as it flies off. Luckily it often lands within view on a rock face or outcrop. It feeds, primarily in the morning and late afternoon, in spinifex and shrubs or on gully floors between rocks. When not feeding it generally retires to shady crevices and ledges where it is difficult to find although it flushes readily, particularly if approached from above.

BANDED FRUIT-DOVE

Kakadu

Another Territory endemic, the Banded Fruit-Dove lives primarily in monsoon forest and feeds on fruit. It can be seen at the base or top of the Arnhem Land escarpment or its outlying remnants. It is a shy, wary bird and often remains in the canopy of the monsoon forest throughout the day. Especially in the late afternoon, it can be seen flying between patches of forest or up and down the escarpment. At times it drinks from pools or soaks which are worth waiting around, particularly in the morning.

VARIED LORIKEET

Darwin

Small Parks of the Top End

Kakadu

Litchfield

The Varied Lorikeet is a mobile woodland species and is often absent for many months until it appears in large numbers. It is usually seen flying overhead in fast, tight and noisy flocks. It feeds on flowering paperbarks, grevilleas and eucalypts where, being predominantly green, it can be difficult to observe in the foliage. It takes flight easily when feeding.

PRINCESS PARROT

Uluru

The nomadic and rare Princess Parrot is considered to be threatened, but is insufficiently known. Although a few sightings have been recorded at Uluru, Watarrka and the West MacDonnells, its stronghold appears to be the sandy deserts to the west of these areas.

NORTHERN ROSELLA

Kakadu

Litchfield

The Northern Rosella is widespread in the Top End but uncommon. Usually seen in pairs or groups of three to five in woodland habitat, it is often flushed from the sides of roads, particularly in the Litchfield region. Its silhouette and undulating flight are typical of rosellas and are distinctive in the Top End.

HOODED PARROT

Katherine

A rare woodland species, the Hooded Parrot is endemic to the Northern Territory where its range is centred around the Katherine region. It is usually seen in small groups and prefers open woodland with a grassy understorey in rolling hill country. The best place to look for it is at waterholes in the early morning and late afternoon when, like other seed eaters, it needs to drink. It comes in to water in noisy groups and after drinking it moves quietly to feeding grounds. During the hotter part of the day it sits quietly in the dense foliage of eucalypts. Patch burning seems to favour this species so it is worth looking in, or near, recently burnt areas.

BOURKE'S PARROT

East MacDonnell Ranges

West MacDonnell Ranges

South East Desert Loop

Bourke's Parrot is a quiet bird primarily of acacia shrubland. It is also often found on gibber plains where it feeds around shrubs and grasses in the early morning. During the hotter parts of the day it is often found in small bushes or in Red Mulga creeklines. When roosting, the birds are quiet and well camouflaged. Just after sunset they travel to roost, at which time they are noisy and active but, because of fading light, difficult to observe.

NIGHT PARROT

The Night Parrot was considered extinct until 1990. It is now considered to be possibly threatened but not enough is known to be certain of its status. It was primarily widespread in inland Australia and there have been unverified sightings in the last decade in the Territory. The latest information indicates that this bird uses chenopod shrublands in dry years and moves into areas of seeding spinifex after rains.

ORIENTAL CUCKOO

Small Parks of the Top End

Kakadu

A wet season migrant, the Oriental Cuckoo is most often seen in the canopy of monsoon forests where it is wary and elusive. It is harassed by other birds in flight. Its flight resembles that of the Australian Hobby. If seen well it can't be confused with other species.

RUFOUS OWL

Small Parks of the Top End

Kakadu

The Rufous Owl is the second largest owl in Australia. Roosting in dense foliage in the canopy of monsoon forest it is very difficult to find. The bird hunts in surrounding woodland and perhaps along the forest edge. Look for it around dusk when it becomes active and starts calling. It is highly territorial and responds well to playbacks of its call. It has been known to roost in the Darwin Botanic Gardens.

LARGE-TAILED NIGHTJAR

Darwin

Small Parks of the Top End

The Large-tailed Nightjar is best seen in the early evening flying over the edge of clearings next to monsoon forest or sometimes over roads near riverine vegetation. It is rarely flushed from the leaf littered floor of its monsoonal habitat during the day. Its white tail markings are distinctive.

LITTLE KINGFISHER

Darwin

Kakadu

The Little Kingfisher is a tiny bird closely associated with fresh or brackish waterways and is difficult to find. Usually seen perched immediately over water in the vegetation, it sits motionless and occasionally bobs its head. Flashes of blue over water can betray its presence.

COLLARED KINGFISHER

Darwin

A large bird for its genus, the Collared Kingfisher is common and readily seen in mangroves, particularly where they are next to mudflats. The bird feeds on exposed mud at low tide and is easy to observe. It is vocal at dawn and dusk. Watch for the similar Sacred Kingfisher that uses the same habitat. The Collared is darker above with clean white underparts and is notably larger, particularly the bill.

RAINBOW PITTA

Small Parks of the Top End

Kakadu

The beautiful Rainbow Pitta is solely a bird of the monsoon forest and generally inhabits patches larger than one hectare. Usually silent and unobtrusive, often the only clue to its presence is a gentle rustling in the leaf litter as it searches for prey. It prefers an open understorey dominated by leaf litter. Birds call from low perches at dawn and dusk and in the early breeding season (the beginning of the wet) they become more vocal.

BLACK-TAILED TREECREEPER

Kakadu

Litchfield

Katherine

The Black-tailed Treecreeper is the only treecreeper of northern Australia and is not as common as its southern relatives. It is patchily distributed, being found in open woodland or hilly country with rough-barked eucalypt trees. Its loud distinctive call is similar to southern species and is the best indication of its presence. It probes the bark of eucalypts for prey and also feeds on the ground.

PURPLE-CROWNED FAIRY-WREN

Victoria River District

The Purple-crowned Fairy-wren is a declining species in the west and is considered vulnerable nationally. It lives around permanent fresh water either in dense pandanus or long grass, often in association with large paperbarks. At times it forages in trees, for instance in the crowns of pandanus or in the foliage of eucalypts, acacias and paperbarks. Generally shy and elusive, this bird has distinctive markings and a more metallic call than other fairy-wrens.

RUFOUS-CROWNED EMU-WREN

West MacDonnell Ranges

The tiny Rufous-crowned Emu-wren lives only in mature spinifex on a variety of landforms from rocky slopes to sand dunes. The dynamics of the population are explosive and the species can be common in one season and virtually nonexistent the next. It is inconspicuous in its habits and its call gives the best indication of its presence. Once disturbed it usually dives into the nearest spinifex tussock where it waits for the danger to pass before climbing up the nearest branch or spine and resuming its previous activity. It looks similar to female and eclipse-plumaged fairy-wrens, but is more rufous, the tail is very sparse and its behaviour is quite different.

WHITE-THROATED GRASSWREN

Kakadu

The striking White-throated Grasswren is endemic to the Northern Territory and is found only on top of the Arnhem Land escarpment in association with spinifex in shallow gullies. This habitat is usually found between the escarpment face and woodland plateau behind. The bird works its way slowly along the gullies foraging in and around spinifex bushes for insects. It is shy and the best way to find it is to listen for its unusual call — a collection of haunting metallic notes and trills that rise and fall and carry some distance. The species is thought to be common but is not often recorded because of its retiring nature and remote habitat.

CARPENTARIAN GRASSWREN

Top End to the Centre

The Carpentarian Grasswren has a restricted range in spinifex-clad sandstone country inland from the Gulf of Carpentaria. It prefers mature spinifex that has not been burnt for a long time, dissected sandstone slopes and escarpments.

EYREAN GRASSWREN

South East Desert Loop

The Eyrean Grasswren solely inhabits the remote Simpson Desert where it lives on the tops of sand dunes dominated by Sandhill Cane-grass and some spinifex. Known from one record in the late 1800s, it was rediscovered in 1976 in South Australia and subsequently recorded in the Northern Territory in 1985. Although its numbers fluctuate considerably it can be common and is always worth searching for. A very shy and retiring bird, it is difficult to observe. Listen for its faint call and look ahead for it scampering from clump to clump. It tends to travel in small groups which, if you split them up, will attempt to regroup.

DUSKY GRASSWREN

East MacDonnell Ranges

West MacDonnell Ranges

Watarrka

The Dusky Grasswren is found on the spinifex-clad ranges of the Centre. It is common within its habitat and can be found at most gorges. It is usually located first by its call, a loud metallic trill. Often seen skipping from rock to rock in small parties, it is easily identified as it is the only grasswren in its habitat.

GREEN-BACKED GERYGONE

Darwin

The Green-backed Gerygone inhabits monsoon forests, fringing mangroves and riverine and billabong habitats of the coastal Top End. Common in broad-leafed coastal vegetation in association with the Yellow White-eye, it has a typical gerygone call and is easily found and approached as it feeds from outer foliage in a hovering fashion.

SLATY-BACKED THORNBILL

Top End to the Centre

Living in the Mulga shrublands, the Slaty-backed Thornbill is usually found alone or in pairs feeding in the outer foliage, and is often associated with Chestnut-rumped and Inland Thornbills. It is uncommon and hard to identify as it resembles the Chestnut-rumped species. The Slaty-backed has a white breast, the forehead is streaked black and the eye is red.

BANDED WHITEFACE

East MacDonnell Ranges

South East Desert Loop

The Banded Whiteface is nomadic on sparsely-vegetated plains and in sand dune country. Listen for its call which is similar to that of the Southern Whiteface but more melodic. It will be found on the ground feeding in the grasses. Watch for it at the edge of gibber plains fringed by gidgee trees or perched on dead trees emerging from living vegetation in interdunal swales. At times it joins mixed flocks, mostly with Southern Whiteface.

HELMETED FRIARBIRD

Darwin

Kakadu

Three races of the Helmeted Friarbird occur in the Northern Territory. One is common in the coastal areas of Darwin where its primary habitat is mangroves and nearby monsoon forest and gardens. The sandstone race inhabits gallery monsoon forest and nearby woodlands of the Arnhem Land escarpment. The third occurs on Melville Island. Silver-crowned Friarbird can be seen in the same habitats and is difficult to distinguish unless observed closely. The Helmeted has a noticeably smaller knob, is larger, more uniformly dark and lacks the silver crown. It is shy and difficult to see well.

WHITE-LINED HONEYEATER

Kakadu

The White-lined Honeyeater is primarily a bird of the escarpment gullies and sides but it will venture into plateau woodland nearby. When eucalypts (especially Woollybutt) and paperbarks are in flower it congregates with other honeyeaters amid much activity and squabbling. At other times it can be found in smaller numbers in pockets of gallery monsoon forest and woodland trees. Its call is characteristic and a reliable guide to its presence.

BAR-BREASTED HONEYEATER

Small Parks of the Top End

Litchfield

Katherine

A nectar dependent bird, the Bar-breasted Honeyeater is often observed at flowering shrubs and trees such as paperbarks, eucalypts and grevilleas along creeklines. It is often seen in association with other honeyeaters near water.

GREY HONEYEATER

Alice Springs

West MacDonnell Ranges

Uluru

A little known bird of the Mulga shrublands, the rare Grey Honeyeater feeds mainly on insects which it gleans from foliage in a way that is unusual for honeyeaters. Reportedly, it also feeds on mistletoe fruits and nectar from eremophilas. Its lack of distinctive features makes it difficult to identify. It associates with thornbills and Western Gerygone.

BLACK HONEYEATER

Alice Springs

East MacDonnell Ranges

West MacDonnell Ranges

The small, highly nomadic Black Honeyeater is widely distributed over inland Australia but is not often observed. It arrives in large numbers after good rains at any time of the year, often becoming the dominant honeyeater. A very active bird, it feeds on flowering plants, particularly eremophilas, which bloom after rains. It also seeks insects attracted to the flowers. It calls from the tops of shrubs or small trees.

PIED HONEYEATER

Alice Springs

East MacDonnell Ranges

West MacDonnell Ranges

Another nomadic bird, the Pied Honeyeater has similar habits and movements to the Black Honeyeater with which it often associates.

YELLOW CHAT

Kakadu

Top End to the Centre

Closely associated with water, both in the major floodplains of the Top End and at bore mounds in the arid regions, the rare Yellow Chat prefers dense reeds. It resides in the Top End and is nomadic in the Centre. In the dry in the Top End it concentrates around more permanent waterways.

GIBBERBIRD

South East Desert Loop

The Gibberbird survives in the extreme conditions of the gibber plains. Gaining moisture from morning dews and insects, and seeking refuge in burrows, it actually flourishes in this environment. It often stands with its back to the observer looking over its shoulder at which time it is superbly camouflaged. When flushed from the side of the road it will fly out to a perch on a vantage point such as a rock, shrub or pile of dirt. Also look for it in vegetated shallow depressions or along small drainage lines on the gibber plain. Richard's Pipit is common in the habitat and is often flushed in the same manner. The Gibberbird has more rounded wings, no white edge to its tail, a chat-like flight and does not flush high into the air.

MANGROVE ROBIN

Darwin

The Mangrove Robin prefers the dense stunted mangroves to landward. This is not an easy habitat to observe and as the bird is uncommon and unobtrusive, it is not often recorded. Although quiet it is not shy and can be approached closely as it clings to the trunk or branch of a mangrove. From here it drops to the mud to seize small crabs and other prey.

WHITE-BROWED ROBIN

Kakadu

Katherine

Found most often in monsoon forest and bamboo thickets along creeklines, the unassuming White-browed Robin uses the mid-storey vegetation and undergrowth areas. It is typically "robin-like" in that it perches on the sides of trunks, bare branches and vines, and also feeds on the ground. Pairs or individuals are usually found evenly spaced along creeks. The territorial call will help locate them.

CHIMING WEDGEBILL

Uluru

South East Desert Loop

The Chiming Wedgebill's presence is given away by its wonderful call, often (and so accurately) described as 'did you get drunk?' It usually calls from the tops of bushes in its habitats of shrubland, plains and sand dunes. A shy bird, it tends to keep just out of sight. Establish its location then attempt to get closer by moving slowly and quietly. It can be confused with the female Crested Bellbird but the calls differ. Its range meets that of the Chirruping Wedgebill in the Simpson Desert but while they are very similar looking their calls are vastly different.

MANGROVE GOLDEN WHISTLER

Darwin

The Mangrove Golden Whistler prefers taller waterside stands of mangroves but will also use the stunted trees on the landward side. It feeds in the mid-storey and at times at the edge along waterways. During the wet it sometimes moves into nearby areas of monsoon forest. Its call is almost identical to that of the common Golden Whistler and is the best means of finding it. Its plumage is also very similar to that of its southern relative. It may be a common bird but due to its inaccessible habitat it is not easy to see.

WHITE-BREASTED WHISTLER

Darwin

The White-breasted Whistler lives in dense stunted mangroves on the landward side and is difficult to approach. It forages in the foliage and branches and drops to the ground to feed on crustaceans. Often in pairs, it is usually located by its call which is similar to that of other whistlers.

SANDSTONE SHRIKE-THRUSH

Kakadu

Victoria River District

The Sandstone Shrike-thrush lives on the escarpment and its haunting and melodic call is characteristic of the habitat. It is most often seen calling from the top of a rocky outcrop with its head pointed skywards. At other times it can be seen chasing insects on the ground in vegetation in rock crevices or hopping over rock faces. The Grey Shrike-thrush prefers woodland habitat, differs in colour and has a distinct white patch on the lores (the space between the eye and the bill).

BROAD-BILLED FLYCATCHER

Darwin

Small Parks of the Top End

Found singly or in pairs, the Broad-Billed Flycatcher mainly inhabits mangroves and monsoon forest but will also use surrounding well-vegetated woodland. Difficult to distinguish from the female of the more common Leaden Flycatcher, the Broad-billed has a distinctive wedge-shaped bill which is unmistakable when seen clearly from underneath. The Leaden has a preference for drier woodland habitats but the two species can occur together.

MANGROVE GREY FANTAIL

Darwin

A bird solely of the mangroves, the Mangrove Grey Fantail is a typical fantail in its behaviour, actively gleaning and chasing insects from ground level to the canopy. It is similar to the Grey Fantail which is rare in the Top End and almost never enters mangroves. In poor light it can be confused with the Rufous Fantail. It is not shy, often calls and can be approached easily.

GROUND CUCKOO-SHRIKE

Top End to the Centre

East MacDonnell Ranges

South East Desert Loop

Highly nomadic, the Ground Cuckoo-shrike can show up anywhere. It prefers open grassland plains and open shrubland and often follows well-wooded creeks and rivers. It is a ground dweller, feeding on insects, usually in groups of four to seven. It is often seen when flushed from the roadside when it tends to fly for several hundred metres before landing. Being a nervous bird it is difficult to approach. In flight its black wings and tail stand out against the white back. This is a distinctive feature along with its high-pitched call. It flies with a slightly heavy tail and glides on downswept wings.

WESTERN BOWERBIRD

West MacDonnell Ranges
Watarrka
Uluru

A harsh loud scolding noise and mimicry will alert you to the presence of the Western Bowerbird. It is common in riverine and woodland habitats and easily identified by its obvious golden spots and iridescent pink crest on the back of its neck. It is resident in the carpark and camping ground at Ormiston Gorge.

STAR FINCH

Victoria River District

Although it has declined in the east, the Star Finch is still locally common but patchy in the western part of its range. Its preferred habitat is tall grasses, rushes and other vegetation along watercourses and swamps, often near woodland. It is usually seen in small to medium-sized flocks, frequently with other finches.

PAINTED FINCH

East MacDonnell Ranges
West MacDonnell Ranges
Watarrka

A beautiful member of the firetail group, the Painted Finch inhabits rocky spinifex-covered hills and ranges. It is typically seen in silhouette while perched on vantage points of rocky outcrops. It is well camouflaged as it feeds among the rocks but its "machine-gun" call gives it away. Its numbers fluctuate and it disperses after good rains and concentrates around waterholes as the environment dries out.

YELLOW-RUMPED MANNIKIN

Victoria River District

The Yellow-rumped Mannikin prefers waterways in open woodland with tall grasses and also occurs on nearby grassy plains. It is possibly threatened but is poorly known.

It is similar in shape and habits to the Chestnut-breasted Mannikin with which it interbreeds and a proportion of the population are hybrids between the two species. The two are often found in mixed flocks and juveniles are virtually indistinguishable. The Yellow-rumped is local and patchy in distribution although more common in the west.

PICTORELLA MANNIKIN

Victoria River District

Top End to the Centre

Yet another Australian finch that is in decline, the Pictorella Mannikin uses drier habitats than other mannikins. It is local in occurrence but can be nomadic and will join large mixed flocks of other finches. In the wet it tends to move southwards to breed.

GOULDIAN FINCH

Katherine

Victoria River District

The spectacularly coloured Gouldian Finch is an endangered species and is apparently still in decline, although the causes of this are not clear. Like many finches of the Top End it feeds on the seeds of tall sorghum grasses in open woodland dominated by Salmon Gum. It may be seen clinging to the seed heads but often feeds on the ground when seed has dropped. It mixes readily with other finches so observers need to look carefully at any large flock. Early morning at an undisturbed waterhole in the dry season is the best opportunity to see it. The birds tend to land in nearby trees or shrubs, move closer in stages and fly down in small groups to drink. Breeding is confined to open woodland with a grassy understorey. It is not well known where these birds go in the wet season.

SPINIFEXBIRD

East MacDonnell Ranges

West MacDonnell Ranges

The Spinifexbird is seen all year round in mature spinifex with a low shrub overstorey. Usually found along small creeks in stony areas on plains, hills and ranges it is recognised by its rufous crown, cocked tail and clean white breast. Often seen alone or in pairs, the first sign of its presence is either its loud alarm or pleasant chirring call. Once disturbed it tends to shelter in a spinifex tussock until curiosity gets the better of it, when it will climb up a nearby shrub or spinifex spine. The most accessible place to see it is at the Ochre Pits in the West MacDonnells.

ZITTING CISTICOLA

Darwin

Kakadu

The Zitting Cisticola occupies saline swampy grasslands near the coast and is not an easy bird to find or identify. It is very similar to the Golden-headed Cisticola and may use similar habitats, although the Golden-headed prefers freshwater grasslands. In the breeding season (the wet) its incessant "tick" call indicates its presence and it will climb grass stems to view the observer. In comparison to the Golden-headed, the Zitting appears more heavily streaked and darker, flies further when flushed and more readily lands on the ground. In the dry the species is silent and almost impossible to find.

ohnny Estbergs

Chestnut-quilled Rock-Pigeon
A Northern Territory endemic, this locally common bird is well camouflaged in its escarpment habitat

Rainbow Pitta
This beautiful bird lives exclusively in monsoon forest. Often the only clue to its presence is a gentle rustling in the leaf litter as it searches for prey

Hooded Parrot (F)
The rare Hooded Parrot is endemic to the Northern Territory. It prefers open woodland with grassy understorey in rolling hill country

Johnny Estbergs

Hooded Parrot (M)

Ian Morris

Gouldian Finch

A nationally endangered species, this finch has its stronghold in the Northern Territory

White-lined Honeyeater
An active and aggressive bird of the Arnhem Land escarpment, its characteristic call alerts birdwatchers to its presence

Ian Morris

Ian Morris

Banded Fruit-Dove
The shy and wary Banded Fruit-Dove lives primarily in monsoon forest

Ian Morris

White-throated Grasswren

The striking White-throated Grasswren is found on top of the Arnhem Land escarpment where it prefers shallow gullies

Partridge Pigeon

An uncommon and patchily distributed inhabitant of Top End woodland, the Partridge Pigeon is often seen on the sides of roads

Raoul Slater

Spinifex Pigeon
The beautiful colours of the Spinifex Pigeon are a perfect camouflage against rusty sandstone and quartzite ranges

NTTC

Riverine habitat provides the greatest bird diversity in the Centre

NTTC

Katherine Gorge, Nitmiluk National Park
Apart from its sheer beauty, Katherine Gorge contains excellent bird life and good walking trails

NTTC

Yellow Water
The most famous wetland in Kakadu is one of the best places in Australia to view waterbirds

Peter Wilkins

West MacDonnell National Park
The harsh environment of the Central Australian ranges provides specialised habitats for sought after birds

NTTC

Kings Canyon, Watarrka National Park
The spectacular Kings Canyon provides good access to special birds of the ranges

Brolgas on wetlands, Kakadu National Park

Glen Helen, West MacDonnell National Park
The Finke River cuts through the ranges near Glen Helen homestead forming the picturesque Glen Helen Gorge where a permanent waterhole provides an excellent habitat for waterbirds

Great, Intermediate and Cattle Egrets
Kakadu National Park contains magnificent wetland areas where waterbirds gather in large numbers

Tnorala (Gosse Bluff)
The high quality spinifex habitat at Tnorala make it a prime site for spinifex species such as the Rufous-crowned Emu-wren and Spinifexbird

Arnhem Land plateau
The western escarpment extends from Kakadu National Park to the Katherine region. Waterfalls and gorges at the escarpment base support monsoon forests

Uluru Kata-Tjuta National Park
The magnificent Kata-Tjuta are home to a number of common shrubland birds

Magpie Geese, Kakadu National Park
As the dry season sets in, water gradually recedes to the permanent billabongs resulting in a massive concentration of birds

Alice Springs sewage ponds
These permanent ponds house a large number of waterbirds and are notable for rare and vagrant species

Spinifex is a common understorey of shrubland in the Centre and an important bird habitat

West MacDonnell National Park
Shrubland, the most widespread habitat in the Centre, occurs on a variety of landforms

Environment Centre NT

Barkly Tablelands
The grass plains are dominated by a dense layer of Mitchell grasses

ibber Plain
he red stony gibber plains are usually devoid f vegetation but after good rains they become overed with ephemeral herbs and grasses

Peter Wilkins

Peter Wilkins

Simpson Desert
Large red sand dunes dominated by Sandhill Cane-grass are the home of the Eyrean Grasswren

CHECKLIST

The Checklist includes all land and regular freshwater, coastal and offshore birds of the Northern Territory. The first column is for recording your observations. The birds are listed by their common and scientific names. The Special Birds column indicates that the species is discussed in the Special Birds chapter.

The rest of the table is split into the Top End and the Centre regions. For each region a status, habitat and area section is included.

The status of birds in the Northern Territory is not well known so in this chapter, it is deliberately described in very broad terms. The status indicates to the observer the chance of seeing a bird in a region and whether it is rare there. The status section is blank if the species does not occur in the region. The following abbreviations are used:

C Common
UC Uncommon
R Rare
V Vagrant
Ex Extinct in the Northern Territory

The movements and migration patterns of birds are also poorly known. However, where applicable, an indication of the species' movement is included in the status section. Abbreviations are as follows:

N Nomadic
SM Summer Migrant

The next two sections indicate the habitat and areas where the birds are likely to occur. Birds listed as vagrants may not be given an area.

Some sites with suitable habitats are inaccessible to the public so in some of the areas where birds are listed, a site may not be given.

Checklist			Top End Habitats							Top End Areas						Centre Habitats							Centre Areas					
Ticks	Common names and scientific names	Special Birds	Status	Woodland	Monsoon Forest	Escarpment	Mangroves	Mudflats & Beaches	Wetlands	Darwin	Small Parks	Kakadu	Litchfield	Katherine	VRD	Status	Shrubland	Riverine	Ranges	Sand Dunes	Plains	Wetlands	Alice Springs	East MacDonnells	West MacDonnells	Watarrka	Uluru	South East Desert
	Emu *Dromaius novaehollandiae*		R	●					●		●	●	●	●	●	UC	●	●	●	●	●	●		●	●	●	●	●
	Malleefowl *Leipoa ocellata*															EX												
	Orange-footed Scrubfowl *Megapodius reinwardt*		C		●		●			●	●	●	●															
	Stubble Quail *Coturnix pectoralis*		V	●					●			●				V	●	●		●	●	●					●	●
	Brown Quail *Coturnix ypsilophora*		C	●					●	●	●	●	●	●	●													
	King Quail *Coturnix chinensis*	✧	UC	●					●	●	●	●	●	●	●													
	Magpie Goose *Anseranas semipalmata*		C						●	●	●	●	●	●	●	V						●	●					
	Plumed Whistling-Duck *Dendrocygna eytoni*		C						●	●	●	●	●	●	●	V						●	●					
	Wandering Whistling-Duck *Dendrocygna arcuata*		C						●	●	●	●	●	●	●	V						●	●					
	Blue-billed Duck *Oxyura australis*		V						●							V						●	●					
	Freckled Duck *Stictonetta naevosa*		V						●			●				V						●	●					
	Black Swan *Cygnus atratus*		UC						●	●	●	●				UC						●	●		●		●	●
	Australian Shelduck *Tadorna tadornoides*															V						●	●					
	Radjah Shelduck *Tadorna radjah*		C				●	●	●	●	●	●	●	●	●													

Australian Wood Duck *Chenonetta jubata*	UC				●	●	●				●	UC	●	●	●	●	●	●	●
Green Pygmy-goose *Nettapus pulchellus*	C				●	●	●	●	●	●	●								
Pacific Black Duck *Anas superciliosa*	C				●	●	●	●	●	●	●	C	●	●	●	●	●	●	●
Australasian Shoveler *Anas rhynchotis*												V	●	●					
Grey Teal *Anas gracilis*	C				●	●	●	●	●	●	●	C	●	●	●	●	●	●	●
Chestnut Teal *Anas castanea*	V				●							V	●	●					
Garganey *Anas querquedula* ✧	RSM				●	●	●	●				V	●	●					
Pink-eared Duck *Malacorhynchus membranaceus*	UC				●	●	●	●		●	●	UC	●	●	●	●	●	●	●
Hardhead *Aythya australis*	UC				●	●	●	●		●	●	UC	●	●	●	●	●	●	●
Australasian Grebe *Tachybaptus novaehollandiae*	C				●	●	●	●	●	●	●	C	●	●	●	●	●	●	●
Hoary-headed Grebe *Poliocephalus poliocephalus*	R				●	●		●				C	●	●	●	●	●	●	●
Great Crested Grebe *Podiceps cristatus*	V				●							V	●	●				●	
Tahiti Petrel *Pseudobulweria rostrata*	V			●															
Streaked Shearwater *Calonectris leucomelas*	V			●															
Wedge-tailed Shearwater *Puffinus pacificus*	V			●															
Hutton's Shearwater *Puffinus huttoni*	V			●															
Brown Booby *Sula leucogaster*	UC			●		●		●											
Darter *Anhinga melanogaster*	C	●	●		●	●	●	●	●	●	●	R	●	●	●	●	●	●	●

Checklist			Top End Habitats							Top End Areas						Centre Habitats							Centre Areas					
Ticks	Common names and scientific names	Special Birds	Status	Woodland	Monsoon Forest	Escarpment	Mangroves	Mudflats & Beaches	Wetlands	Darwin	Small Parks	Kakadu	Litchfield	Katherine	VRD	Status	Shrubland	Riverine	Ranges	Sand Dunes	Plains	Wetlands	Alice Springs	East MacDonnells	West MacDonnells	Watarrka	Uluru	South East Desert
	Little Pied Cormorant *Phalacrocorax melanoleucos*		C			●	●	●	●	●	●	●	●	●	●	UC						●	●	●	●		●	●
	Pied Cormorant *Phalacrocorax varius*		UC			●	●	●	●	●	●	●	●	●	●	UC						●	●	●	●		●	●
	Little Black Cormorant *Phalacrocorax sulcirostris*		C			●	●	●	●	●	●	●	●	●	●	UC						●	●	●	●		●	●
	Great Cormorant *Phalacrocorax carbo*		UC						●	●	●	●		●	●	R						●	●	●	●	●	●	●
	Australian Pelican *Pelecanus conspicillatus*		UC					●	●	●	●	●	●	●	●	R						●	●		●		●	●
	Great Frigatebird *Fregata minor*		RSM					●		●		●																
	Lesser Frigatebird *Fregata ariel*		RSM					●		●		●																
	White-faced Heron *Egretta novaehollandiae*		C	●		●	●	●	●	●	●	●	●	●	●	C						●	●	●	●	●	●	●
	Little Egret *Egretta garzetta*		C	●		●	●	●	●	●	●	●	●	●	●	R						●	●		●	●	●	
	Eastern Reef Egret *Egretta sacra*		C				●	●		●		●			●													
	White-necked Heron *Ardea pacifica*		C	●		●		●	●	●	●	●	●	●	●	UC						●	●	●	●	●	●	●
	Great-billed Heron *Ardea sumatrana*	✧	UC				●	●	●	●	●	●	●	●	●													
	Pied Heron *Ardea picata*		C				●	●	●	●	●	●	●	●	●													
	Great Egret *Ardea alba*		C	●			●	●	●	●	●	●	●	●	●	UC						●	●	●	●	●	●	●

Intermediate Egret *Ardea intermedia*		C				●	●	●	●	●	●	●	●	●	UC						●	●		●			
Cattle Egret *Ardea ibis*		C	●					●	●	●	●		●	●	V						●	●					
Striated Heron *Butorides striatus*		C				●	●		●		●			●	V						●	●					
Nankeen Night Heron *Nycticorax caledonicus*		C		●	●	●	●	●	●	●	●	●	●	●	R						●	●	●	●	●	●	●
Little Bittern *Ixobrychus minutus*		V						●																			
Black Bittern *Ixobrychus flavicollis*	✧	UC			●	●		●	●	●	●	●	●	●													
Glossy Ibis *Plegadis falcinellus*		UC						●	●	●	●	●	●	●	R						●	●		●		●	●
Australian White Ibis *Threskiornis molucca*		C	●			●	●	●	●	●	●	●	●	●	R						●	●				●	
Straw-necked Ibis *Threskiornis spinicollis*		C	●					●	●	●	●	●	●	●	R						●	●		●		●	
Royal Spoonbill *Platalea regia*		UC					●	●	●	●	●	●	●	●	R						●	●		●	●	●	●
Yellow-billed Spoonbill *Platalea flavipes*		UC				●	●	●	●	●	●	●	●	●	R						●	●		●		●	●
Black-necked Stork *Ephippiorhynchus asiaticus*		C				●	●	●	●	●	●	●	●	●	V						●	●					
Osprey *Pandion haliaetus*		UC				●	●	●	●	●	●		●	●	V		●				●			●			
Pacific Baza *Aviceda subcristata*		UC	●	●	●				●	●	●	●	●	●													
Black-shouldered Kite *Elanus axillaris*		UC	●					●	●	●	●	●	●	●	UC		●				●	●	●	●	●	●	●
Letter-winged Kite *Elanus scriptus*	✧	R	●					●	●	●	●				V	●	●			●	●						●
Square-tailed Kite *Lophoictinia isura*		R	●						●	●	●	●	●	●	V	●	●			●	●	●		●		●	●
Black-breasted Buzzard *Hamirostra melanosternon*		UC	●						●	●	●	●	●	●	UC	●	●	●		●		●	●	●	●	●	●

Checklist			Top End Habitats							Top End Areas						Centre Habitats							Centre Areas					
Ticks	Common names and scientific names	Special Birds	Status	Woodland	Monsoon Forest	Escarpment	Mangroves	Mudflats & Beaches	Wetlands	Darwin	Small Parks	Kakadu	Litchfield	Katherine	VRD	Status	Shrubland	Riverine	Ranges	Sand Dunes	Plains	Wetlands	Alice Springs	East MacDonnells	West MacDonnells	Watarrka	Uluru	South East Desert
	Black Kite *Milvus migrans*		C	●	●	●	●	●	●	●	●	●	●	●	●	C	●	●	●	●	●	●	●	●	●	●	●	●
	Whistling Kite *Haliastur sphenurus*		C	●	●	●	●	●	●	●	●	●	●	●	●	C	●	●	●	●	●	●	●	●	●	●	●	●
	Brahminy Kite *Haliastur indus*		C	●			●	●	●	●	●	●	●		●													
	White-bellied Sea-Eagle *Haliaeetus leucogaster*		C			●	●	●	●	●	●	●	●	●	●													
	Spotted Harrier *Circus assimilis*		R	●					●	●	●	●	●	●	●	UC	●	●	●	●	●	●	●	●	●	●	●	●
	Swamp Harrier *Circus approximans*		R						●	●	●	●	●		●	R						●	●					●
	Brown Goshawk *Accipiter fasciatus*		C	●		●			●	●	●	●	●	●	●	UC	●	●			●	●	●	●	●	●	●	●
	Grey Goshawk *Accipiter novaehollandiae*		UC	●	●	●			●	●	●	●	●	●	●													
	Collared Sparrowhawk *Accipiter cirrhocephalus*		C	●	●	●				●	●	●	●	●	●	UC	●	●			●	●	●	●	●	●	●	●
	Red Goshawk *Erythrotriorchis radiatus*	✧	R	●	●							●	●	●	●	V		●										
	Wedge-tailed Eagle *Aquila audax*		UC	●						●	●	●	●	●	●	C	●	●	●	●	●	●	●	●	●	●	●	●
	Little Eagle *Hieraaetus morphnoides*		UC	●						●	●	●	●	●	●	UC	●	●	●	●	●	●	●	●	●	●	●	●
	Brown Falcon *Falco berigora*		C	●		●			●	●	●	●	●	●	●	C	●	●	●	●	●	●	●	●	●	●	●	●
	Australian Hobby *Falco longipennis*		UC	●		●		●	●	●	●	●	●	●	●	UC	●	●	●			●	●	●	●	●	●	●

Grey Falcon *Falco hypoleucos*	✧	RN	●								●		●	●	R	●	●			●			●	●	●	●	●
Black Falcon *Falco subniger*		UC	●		●			●			●	●	●	●	UC	●	●		●	●	●	●	●	●	●	●	●
Peregrine Falcon *Falco peregrinus*		UC	●		●		●		●	●	●	●	●	●	UC	●	●	●			●	●	●	●	●	●	●
Nankeen Kestrel *Falco cenchroides*		C	●		●			●	●	●	●	●	●	●	C	●	●	●	●	●	●	●	●	●	●	●	●
Sarus Crane *Grus antigone*		V						●			●																
Brolga *Grus rubicunda*		C						●	●	●	●	●	●	●	V					●	●						
Buff-banded Rail *Gallirallus philippensis*		UC	●			●		●	●	●	●	●	●	●	V						●	●		●			
Bush-hen *Amaurornis olivaceus*		R		●	●			●	●		●	●															
Baillon's Crake *Porzana pusilla*		R						●	●	●	●			●	V						●	●		●			
Australian Spotted Crake *Porzana fluminea*		V						●							V						●	●		●			
White-browed Crake *Porzana cinerea*	✧	UC		●				●	●	●	●			●													
Chestnut Rail *Eulabeornis castaneoventris*	✧	UC				●			●	●	●			●													
Purple Swamphen *Porphyrio porphyrio*		R						●		●	●	●	●	●	R						●	●		●		●	
Dusky Moorhen *Gallinula tenebrosa*		V						●							R						●	●		●			
Black-tailed Native-hen *Gallinula ventralis*		V						●						●	UCN						●	●		●	●	●	●
Eurasian Coot *Fulica atra*		C						●	●	●	●	●	●	●	C						●	●	●	●		●	●
Australian Bustard *Ardeotis australis*		UC	●					●		●	●	●	●	●	UC	●				●		●	●	●	●	●	●
Red-backed Button-quail *Turnix maculosa*	✧	UC	●		●			●	●	●	●	●	●														

Checklist			Top End Habitats							Top End Areas					
Ticks	Common names and scientific names	Special Birds	Status	Woodland	Monsoon Forest	Escarpment	Mangroves	Mudflats & Beaches	Wetlands	Darwin	Small Parks	Kakadu	Litchfield	Katherine	VRD
	Little Button-quail *Turnix velox*		V	●								●	●	●	●
	Red-chested Button-quail *Turnix pyrrhothorax*	✧	UC	●					●	●	●	●	●	●	●
	Chestnut-backed Button-quail *Turnix castanota*	✧	R	●		●						●		●	●
	Latham's Snipe *Gallinago hardwickii*		RSM						●	●	●	●			
	Swinhoe's Snipe *Gallinago megala*		RSM						●	●	●	●			
	Black-tailed Godwit *Limosa limosa*		CSM					●		●		●			●
	Bar-tailed Godwit *Limosa lapponica*		CSM					●		●		●			●
	Little Curlew *Numenius minutus*	✧	CSM						●	●	●	●	●	●	●
	Whimbrel *Numenius phaeopus*		CSM				●	●	●	●		●	●		●
	Eastern Curlew *Numenius madagascariensis*		CSM				●	●		●		●			●
	Common Redshank *Tringa totanus*		V					●	●						
	Marsh Sandpiper *Tringa stagnatilis*		CSM				●	●	●	●	●	●		●	●
	Common Greenshank *Tringa nebularia*		CSM				●	●	●	●	●	●	●	●	●
	Wood Sandpiper *Tringa glareola*		UCSM						●	●	●	●		●	●

Checklist	Centre Habitats							Centre Areas					
Common names and scientific names	Status	Shrubland	Riverine	Ranges	Sand Dunes	Plains	Wetlands	Alice Springs	East MacDonnells	West MacDonnells	Watarrka	Uluru	South East Desert
Little Button-quail *Turnix velox*	UC	●	●	●	●	●	●	●	●	●	●	●	●
Red-chested Button-quail *Turnix pyrrhothorax*	V	●				●							
Chestnut-backed Button-quail *Turnix castanota*													
Latham's Snipe *Gallinago hardwickii*													
Swinhoe's Snipe *Gallinago megala*	RSM						●	●					
Black-tailed Godwit *Limosa limosa*	RSM						●	●					
Bar-tailed Godwit *Limosa lapponica*	RSM						●	●					
Little Curlew *Numenius minutus*	RSM						●	●					
Whimbrel *Numenius phaeopus*													
Eastern Curlew *Numenius madagascariensis*													
Common Redshank *Tringa totanus*													
Marsh Sandpiper *Tringa stagnatilis*	UCSM						●	●		●		●	●
Common Greenshank *Tringa nebularia*	UCSM						●	●		●		●	●
Wood Sandpiper *Tringa glareola*	CSM						●	●		●		●	

Terek Sandpiper *Xenus cinereus*		CSM	●	●		●		●			●								
Common Sandpiper *Actitis hypoleucos*		CSM	●	●	●	●	●	●	●	●	●	CSM	●	●		●		●	
Grey-tailed Tattler *Heteroscelus brevipes*		CSM	●	●		●		●			●								
Wandering Tattler *Heteroscelus incanus*		V	●	●		●													
Ruddy Turnstone *Arenaria interpres*		CSM		●		●		●			●	V	●	●					
Asian Dowitcher *Limnodromus semipalmatus*		V		●		●													
Great Knot *Calidris tenuirostris*		CSM		●		●		●			●								
Red Knot *Calidris canutus*		CSM		●		●		●			●								
Sanderling *Calidris alba*		UCSM		●		●		●											
Little Stint *Calidris minuta*		V		●	●														
Red-necked Stint *Calidris ruficollis*		CSM	●	●	●	●	●	●		●	●	UCSM	●	●				●	
Long-toed Stint *Calidris subminuta*		V			●							V	●	●					
Pectoral Sandpiper *Calidris melanotos*		V			●			●				V	●	●					
Sharp-tailed Sandpiper *Calidris acuminata*		CSM	●	●	●	●	●	●			●	UCSM	●	●		●		●	●
Curlew Sandpiper *Calidris ferruginea*		CSM	●	●	●	●	●	●			●	V	●	●					
Broad-billed Sandpiper *Limicola falcinellus*	✧	RSM	●	●	●	●	●	●				V	●	●					
Ruff *Philomachus pugnax*		V			●							V	●	●					
Red-necked Phalarope *Phalaropus lobatus*		V		●															

Checklist			Top End Habitats							Top End Areas						Centre Habitats							Centre Areas					
Ticks	Common names and scientific names	Special Birds	Status	Woodland	Monsoon Forest	Escarpment	Mangroves	Mudflats & Beaches	Wetlands	Darwin	Small Parks	Kakadu	Litchfield	Katherine	VRD	Status	Shrubland	Riverine	Ranges	Sand Dunes	Plains	Wetlands	Alice Springs	East MacDonnells	West MacDonnells	Watarrka	Uluru	South East Desert
	Painted Snipe *Rostratula benghalensis*		V						●							V						●	●					
	Comb-crested Jacana *Irediparra gallinacea*		C						●	●	●	●	●	●	●													
	Bush Stone-curlew *Burhinus grallarius*		C	●		●			●	●	●	●	●	●	●	R	●	●						●	●	●	●	●
	Beach Stone-curlew *Esacus neglectus*	✧	UC					●		●		●			●													
	Pied Oystercatcher *Haematopus longirostris*		UC					●		●		●			●													
	Sooty Oystercatcher *Haematopus fuliginosus*		UC					●		●		●			●													
	Black-winged Stilt *Himantopus himantopus*		C					●	●	●	●	●	●		●	UC						●	●		●		●	●
	Banded Stilt *Cladorhynchus leucocephalus*															V						●	●					
	Red-necked Avocet *Recurvirostra novaehollandiae*		R						●	●	●	●				UC						●	●		●		●	
	Pacific Golden Plover *Pluvialis fulva*		UCSM					●	●	●	●				●	V						●	●					
	Grey Plover *Pluvialis squatarola*		CSM					●		●		●			●	V						●	●					
	Ringed Plover *Charadrius hiaticula*		V					●				●																
	Little Ringed Plover *Charadrius dubius*		V					●				●																
	Red-capped Plover *Charadrius ruficapillus*		C					●	●	●	●	●	●	●	●	UC						●	●		●		●	

Lesser Sand Plover *Charadrius mongolus*		CSM					●	●	●		●				V						●	●					●
Greater Sand Plover *Charadrius leschenaultii*		CSM					●	●	●		●				V						●	●					
Oriental Plover *Charadrius veredus*	✧	RSM	●				●	●	●	●	●	●	●	●	RN	●				●	●	●	●	●		●	●
Inland Dotterel *Charadrius australis*															UCN					●	●	●	●			●	●
Black-fronted Dotterel *Elseyornis melanops*		UC						●	●	●	●	●	●	●	C		●				●	●	●	●	●	●	●
Red-kneed Dotterel *Erythrogonys cinctus*		UC						●	●	●	●	●	●	●	UC						●	●	●	●		●	
Banded Lapwing *Vanellus tricolor*															R					●		●	●	●	●	●	●
Masked Lapwing *Vanellus miles*		C	●			●		●	●	●	●	●	●	●	UC		●			●	●	●	●	●	●	●	●
Oriental Pratincole *Glareola maldivarum*	✧	N	●					●	●	●	●	●	●	●	V					●	●	●					
Australian Pratincole *Stiltia isabella*		C	●					●	●	●	●	●	●	●	UC					●	●	●	●		●	●	●
Pomarine Jaeger *Stercorarius pomarinus*		V					●																				
Arctic Jaeger *Stercorarius parasiticus*		V					●																				
Silver Gull *Larus novaehollandiae*		UC				●	●	●	●		●			●	R						●	●					
Gull-billed Tern *Sterna nilotica*		C					●	●	●	●	●		●	●	V						●	●					
Caspian Tern *Sterna caspia*		UC					●	●	●	●	●		●	●	V						●	●					
Lesser Crested Tern *Sterna bengalensis*		UC					●		●		●			●													
Crested Tern *Sterna bergii*		C					●	●	●		●			●													
Roseate Tern *Sterna dougallii*		UC					●		●		●																

Checklist			Top End Habitats							Top End Areas						Centre Habitats							Centre Areas					
Ticks	Common names and scientific names	Special Birds	Status	Woodland	Monsoon Forest	Escarpment	Mangroves	Mudflats & Beaches	Wetlands	Darwin	Small Parks	Kakadu	Litchfield	Katherine	VRD	Status	Shrubland	Riverine	Ranges	Sand Dunes	Plains	Wetlands	Alice Springs	East MacDonnells	West MacDonnells	Watarrka	Uluru	South East Desert
	Black-naped Tern *Sterna sumatrana*		R					●		●		●																
	Common Tern *Sterna hirundo*		R					●		●		●			●													
	Little Tern *Sterna albifrons*		R					●		●		●			●													
	Bridled Tern *Sterna anaethetus*		V					●				●																
	Sooty Tern *Sterna fuscata*		V					●																				
	Whiskered Tern *Chlidonias hybridus*		C					●	●	●	●	●	●	●	●	UC						●	●		●		●	
	White-winged Black Tern *Chlidonias leucopterus*		CSM					●	●	●	●	●		●	●	UCSM						●	●				●	
	Common Noddy *Anous stolidus*		V					●																				
	Rock Dove *Columba livia*		UC							●						R							●				●	
	Emerald Dove *Chalcophaps indica*		C		●	●	●			●	●	●	●	●	●													
	Common Bronzewing *Phaps chalcoptera*		UC	●		●				●	●	●	●	●	●	UC	●	●	●	●	●	●	●	●	●	●	●	●
	Flock Bronzewing *Phaps histrionica*	✧	V							●		●				RN					●	●	●	●				●
	Crested Pigeon *Ocyphaps lophotes*		UC	●						●		●	●	●	●	C	●	●	●	●	●	●	●	●	●	●	●	●
	Spinifex Pigeon *Geophaps plumifera*	✧	R	●		●						●			●	C			●			●	●	●	●	●	●	●

Partridge Pigeon *Geophaps smithii*	✧	UC	●							●	●	●	●													
White-quilled Rock-Pigeon *Petrophassa albipennis*		UC			●								●													
Chestnut-quilled Rock-Pigeon *Petrophassa rufipennis*	✧	C			●					●		●														
Diamond Dove *Geopelia cuneata*		UC	●		●			●	●	●	●	●	●	UC	●	●	●	●	●	●	●	●	●	●	●	●
Peaceful Dove *Geopelia striata*		C	●		●		●	●	●	●	●	●	●	UC	●	●	●	●	●	●	●	●	●	●	●	●
Bar-shouldered Dove *Geopelia humeralis*		C	●	●	●	●	●	●	●	●	●	●	●													
Banded Fruit-Dove *Ptilinopus cinctus*	✧	UC		●	●					●																
Rose-crowned Fruit-Dove *Ptilinopus regina*		C		●				●	●	●	●	●														
Pied Imperial-Pigeon *Ducula bicolor*		C	●	●	●	●		●	●	●	●	●	●													
Red-tailed Black-Cockatoo *Calyptorhynchus banskii*		C	●		●		●	●	●	●	●	●	●	UC	●	●	●	●	●	●	●	●	●	●	●	●
Galah *Cacatua roseicapilla*		C	●		●		●	●	●	●	●	●	●	UC	●	●	●	●	●	●	●	●	●	●	●	●
Little Corella *Cacatua sanguinea*		C	●				●	●	●	●	●	●	●	UC	●	●			●	●	●				●	●
Major Mitchell's Cockatoo *Cacatua leadbeateri*														UC	●	●		●	●			●	●	●	●	●
Sulphur-crested Cockatoo *Cacatua galerita*		C	●	●	●	●	●	●	●	●	●	●	●													
Cockatiel *Nymphicus hollandicus*		UC	●		●			●	●	●	●	●	●	CN	●	●	●	●	●	●	●	●	●	●	●	●
Rainbow Lorikeet *Trichoglossus haematodus*		C	●	●	●	●	●	●	●	●	●	●	●													
Varied Lorikeet *Psitteuteles versicolor*	✧	N	●		●	●	●	●	●	●	●	●	●													
Red-winged Parrot *Aprosmictus erythropterus*		C	●	●	●		●	●	●	●	●	●	●													

Checklist			Top End Habitats							Top End Areas						Centre Habitats							Centre Areas					
Ticks	Common names and scientific names	Special Birds	Status	Woodland	Monsoon Forest	Escarpment	Mangroves	Mudflats & Beaches	Wetlands	Darwin	Small Parks	Kakadu	Litchfield	Katherine	VRD	Status	Shrubland	Riverine	Ranges	Sand Dunes	Plains	Wetlands	Alice Springs	East MacDonnells	West MacDonnells	Watarrka	Uluru	South East Desert
	Princess Parrot *Polytelis alexandrae*	✧														RN		●		●	●				●	●	●	
	Northern Rosella *Platycercus venustus*	✧	UC	●		●				●	●	●	●	●	●													
	Australian Ringneck *Barnardius zonarius*															C	●	●	●			●	●	●	●	●	●	●
	Red-rumped Parrot *Psephotus haematonotus*															V		●										●
	Mulga Parrot *Psephotus varius*															UC	●	●			●		●	●	●	●	●	●
	Hooded Parrot *Psephotus dissimilis*	✧	R	●		●						●		●														
	Budgerigar *Melopsittacus undulatus*		N	●								●		●	●	CN	●	●	●	●	●	●	●	●	●	●	●	●
	Bourke's Parrot *Neopsephotus bourkii*	✧														R	●	●			●			●	●	●	●	●
	Scarlet-chested Parrot *Neophema splendida*															RN	●			●	●						●	
	Night Parrot *Pezoporus occidentalis*	✧	?												?	R	●		●	●	●							
	Oriental Cuckoo *Cuculus saturatus*	✧	RSM	●	●	●	●		●	●	●	●	●	●	●													
	Pallid Cuckoo *Cuculus pallidus*		UCSM	●	●	●			●	●	●	●	●	●	●	UC	●	●	●	●	●	●	●	●	●	●	●	●
	Brush Cuckoo *Cacomantis variolosus*		CSM	●	●	●	●		●	●	●	●	●	●	●													
	Black-eared Cuckoo *Chrysococcyx osculans*		R	●	●	●			●	●	●	●		●	●	R	●		●				●	●	●		●	●

Horsfield's Bronze-Cuckoo *Chrysococcyx basalis*		UC	●		●			●	●	●	●	●	●	●	UC	●	●	●	●	●	●	●	●	●	●	●
Little Bronze-Cuckoo *Chrysococcyx minutillus*		UC		●		●		●	●	●	●	●	●	●												
Common Koel *Eudynamys scolopacea*		CSM	●	●	●	●		●	●	●	●	●	●	●												
Channel-billed Cuckoo *Scythrops novaehollandiae*		UCSM	●	●	●			●	●	●	●	●	●	●	RSM		●	●			●		●		●	●
Pheasant Coucal *Centropus phasianinus*		C	●	●	●			●	●	●	●	●	●	●												
Rufous Owl *Ninox rufa*	✧	UC	●	●	●				●	●	●	●	●													
Barking Owl *Ninox connivens*		C	●	●	●			●	●	●	●	●	●	●												
Southern Boobook *Ninox novaeseelandiae*		C	●	●	●			●	●	●	●	●	●	●	UC	●	●				●	●	●	●	●	●
Masked Owl *Tyto novaehollandiae*		R	●		●						●															
Barn Owl *Tyto alba*		UC	●					●	●	●	●	●	●	●	R		●	●				●	●	●	●	●
Grass Owl *Tyto capensis*		V						●			●															
Tawny Frogmouth *Podargus strigoides*		C	●	●	●	●		●	●	●	●	●	●	●	UC	●	●	●			●	●	●	●	●	●
Spotted Nightjar *Eurostopodus argus*		C	●		●				●	●	●	●	●	●	CN	●	●	●		●	●	●	●	●	●	●
Large-tailed Nightjar *Caprimulgus macrurus*	✧	UC		●		●		●	●	●	●	●	●	●												
Australian Owlet-nightjar *Aegotheles cristatus*		C	●	●	●				●	●	●	●	●	●	C	●	●	●			●	●	●	●	●	●
White-throated Needletail *Hirundapus caudacutus*		V									●				V						●					
Fork-tailed Swift *Apus pacificus*		UCSM	●	●	●	●	●	●	●	●	●	●	●	●	RSM						●	●	●	●	●	●
Azure Kingfisher *Alcedo azurea*		C		●		●		●	●	●	●	●	●	●												

Checklist			Top End Habitats							Top End Areas						Centre Habitats							Centre Areas					
Ticks	Common names and scientific names	Special Birds	Status	Woodland	Monsoon Forest	Escarpment	Mangroves	Mudflats & Beaches	Wetlands	Darwin	Small Parks	Kakadu	Litchfield	Katherine	VRD	Status	Shrubland	Riverine	Ranges	Sand Dunes	Plains	Wetlands	Alice Springs	East MacDonnells	West MacDonnells	Watarrka	Uluru	South East Desert
	Little Kingfisher *Alcedo pusilla*	✧	UC		●		●		●	●	●	●	●	●														
	Blue-winged Kookaburra *Dacelo leachii*		C	●	●	●			●	●	●	●	●	●	●													
	Forest Kingfisher *Todiramphus macleayii*		C	●	●	●	●		●	●	●	●	●	●	●													
	Red-backed Kingfisher *Todiramphus pyrrhopygia*		UC	●					●	●	●	●	●	●	●	C	●	●	●	●	●	●	●	●	●	●	●	●
	Sacred Kingfisher *Todiramphus sanctus*		C	●	●	●	●		●	●	●	●	●	●	●	UC		●					●	●	●	●	●	
	Collared Kingfisher *Todiramphus chloris*	✧	C				●	●		●		●			●													
	Rainbow Bee-eater *Merops ornatus*		C	●	●	●	●	●	●	●	●	●	●	●	●	CSM	●	●	●			●	●	●	●	●	●	●
	Dollarbird *Eurystomus orientalis*		CSM	●	●	●			●	●	●	●	●	●	●													
	Rainbow Pitta *Pitta iris*	✧	UC		●					●	●	●	●															
	White-browed Treecreeper *Climacteris affinis*															R	●	●	●		●			●	●		●	●
	Black-tailed Treecreeper *Climacteris melanura*	✧	UC	●		●				●	●	●	●	●	●													
	Purple-crowned Fairy-wren *Malurus coronatus*	✧	UC						●						●													
	Splendid Fairy-wren *Malurus splendens*															UC	●						●	●	●	●	●	●
	Variegated Fairy-wren		UC	●		●						●	●	●	●	C	●	●	●	●	●	●	●	●	●	●	●	●

White-winged Fairy-wren *Malurus leucopterus*															UC	●			●		●	●	●	●	●	●	●
Red-backed Fairy-wren *Malurus melanocephalus*		C	●		●			●	●	●	●	●	●	●													
Rufous-crowned Emu-wren *Stipiturus ruficeps*	✧														UC			●	●				●	●	●	●	●
White-throated Grasswren *Amytornis woodwardi*	✧	UC			●						●		●														
Carpentarian Grasswren *Amytornis dorotheae*	✧	R			●																						
Striated Grasswren *Amytornis striatus*															R	●								●		●	●
Eyrean Grasswren *Amytornis goyderi*	✧														UC				●								●
Thick-billed Grasswren *Amytornis textilis*															R					●							●
Dusky Grasswren *Amytornis purnelli*	✧														C			●				●	●	●	●	●	
Red-browed Pardalote *Pardalotus rubricatus*		UC											●	●	C	●	●	●				●	●	●	●	●	●
Striated Pardalote *Pardalotus striatus*		C	●		●				●	●	●	●	●	●	C	●	●	●				●	●	●	●	●	●
Rufous Fieldwren *Calamanthus campestris*															R	●											●
Redthroat *Pyrrholaemus brunneus*															R	●						●	●	●	●	●	●
Weebill *Smicrornis brevirostris*		C	●	●	●				●	●	●	●	●	●	C	●	●					●	●	●	●	●	●
Mangrove Gerygone *Gerygone levigaster*		C				●			●		●			●													
Western Gerygone *Gerygone fusca*		R	●										●	●	UC	●						●	●	●	●	●	●
Large-billed Gerygone *Gerygone magnirostris*		C		●	●	●		●	●	●	●	●	●	●													
Green-backed Gerygone *Gerygone chloronotus*	✧	C		●	●	●		●	●	●	●	●	●	●													

Checklist			Top End Habitats							Top End Areas						Centre Habitats							Centre Areas					
Ticks	Common names and scientific names	Special Birds	Status	Woodland	Monsoon Forest	Escarpment	Mangroves	Mudflats & Beaches	Wetlands	Darwin	Small Parks	Kakadu	Litchfield	Katherine	VRD	Status	Shrubland	Riverine	Ranges	Sand Dunes	Plains	Wetlands	Alice Springs	East MacDonnells	West MacDonnells	Watarrka	Uluru	South East Desert
	White-throated Gerygone *Gerygone olivacea*		C	●						●	●	●	●	●	●													
	Inland Thornbill *Acanthiza apicalis*															UC	●	●	●				●	●	●	●	●	●
	Chestnut-rumped Thornbill *Acanthiza uropygialis*															C	●	●	●				●	●	●	●	●	●
	Slaty-backed Thornbill *Acanthiza robustirostris*	✧														R	●								●		●	●
	Yellow-rumped Thornbill *Acanthiza chrysorrhoa*															UC	●	●	●				●	●	●	●	●	●
	Southern Whiteface *Aphelocephala leucopsis*															C	●		●	●	●		●	●	●	●	●	●
	Banded Whiteface *Aphelocephala nigricincta*	✧														R				●	●			●	●	●	●	●
	Spiny-cheeked Honeyeater *Acanthagenys rufogularis*															C	●	●	●		●	●	●	●	●	●	●	●
	Helmeted Friarbird *Philemon buceroides*	✧	UC		●	●	●			●		●		●	●													
	Silver-crowned Friarbird *Philemon argenticeps*		C	●	●	●	●		●	●	●	●	●	●	●													
	Little Friarbird *Philemon citreogularis*		C	●		●			●	●	●	●	●	●	●													
	Blue-faced Honeyeater *Entomyzon cyanotis*		C	●	●	●			●	●	●	●	●	●	●													
	Yellow-throated Miner *Manorina flavigula*		C	●						●	●	●	●	●	●	C	●	●					●	●	●	●	●	●
	White-lined Honeyeater *Meliphaga albilineata*	✧	UC	●	●	●						●																

Singing Honeyeater *Lichenostomus virescens*	R	•		•						•	•	•	•	UC	•		•				•	•	•	•	•	•
White-gaped Honeyeater *Lichenostomus unicolor*	C	•	•	•	•		•	•	•	•	•	•	•													
Grey-headed Honeyeater *Lichenostomus keartlandi*	R	•		•									•	C	•	•	•				•	•	•	•	•	•
Grey-fronted Honeyeater *Lichenostomus plumulus*	R	•		•									•	R	•							•	•	•	•	•
Yellow-tinted Honeyeater *Lichenostomus flavescens*	C	•								•		•	•													
White-plumed Honeyeater *Lichenostomus penicillatus*														C	•	•					•	•	•	•	•	•
Black-chinned Honeyeater *Melithreptus gularis*	UC	•								•	•	•	•	UC	•	•	•				•	•	•	•		•
White-throated Honeyeater *Melithreptus albogularis*	C	•	•	•			•	•	•	•	•	•	•													
Brown Honeyeater *Lichmera indistincta*	C	•	•	•	•		•	•	•	•	•	•	•	C	•	•	•				•	•	•	•	•	•
White-fronted Honeyeater *Phylidonyris albifrons*														UCN	•						•	•	•	•	•	•
Bar-breasted Honeyeater ✧ *Ramsayornis fasciatus*	N	•		•			•	•	•	•	•	•	•													
Rufous-banded Honeyeater *Conopophila albogularis*	C	•	•	•	•		•	•	•	•	•	•	•													
Rufous-throated Honeyeater *Conopophila rufogularis*	C	•		•			•		•	•	•	•	•													
Grey Honeyeater ✧ *Conopophila whitei*														R	•						•		•		•	
Banded Honeyeater *Certhionyx pectoralis*	UC	•		•				•	•	•	•	•	•													
Black Honeyeater ✧ *Certhionyx niger*	N	•											•	R	•		•		•		•	•	•		•	•
Pied Honeyeater ✧ *Certhionyx variegatus*	N	•											•	R	•		•		•		•	•	•		•	•
Dusky Honeyeater *Myzomela obscura*	C	•	•	•	•		•	•	•	•	•	•	•													

Checklist			Top End Habitats							Top End Areas						Centre Habitats							Centre Areas					
Ticks	Common names and scientific names	Special Birds	Status	Woodland	Monsoon Forest	Escarpment	Mangroves	Mudflats & Beaches	Wetlands	Darwin	Small Parks	Kakadu	Litchfield	Katherine	VRD	Status	Shrubland	Riverine	Ranges	Sand Dunes	Plains	Wetlands	Alice Springs	East MacDonnells	West MacDonnells	Watarrka	Uluru	South East Desert
	Red-headed Honeyeater *Myzomela erythrocephala*		C		●		●		●	●	●	●	●	●	●													
	Crimson Chat *Epthianura tricolor*															CN	●	●	●	●	●	●	●	●	●	●	●	●
	Orange Chat *Epthianura aurifrons*															UCN					●	●	●	●	●	●	●	●
	Yellow Chat *Epthianura crocea*	✧	R						●	●	●	●			●	R						●	●					●
	Gibberbird *Ashbyia lovensis*	✧														R					●							●
	Jacky Winter *Microeca fascinans*		R	●								●	●	●	●	R	●	●					●	●	●		●	●
	Lemon-bellied Flycatcher *Microeca flavigaster*		C	●	●	●	●		●	●	●	●	●	●	●													
	Red-capped Robin *Petroica goodenovii*															C	●						●	●	●	●	●	●
	Hooded Robin *Melanodryas cucullata*		R	●								●	●	●	●	UC	●		●				●	●	●	●	●	●
	Mangrove Robin *Eopsaltria pulverulenta*	✧	UC				●			●		●			●													
	White-browed Robin *Poecilodryas superciliosa*	✧	UC	●	●	●			●		●	●	●	●	●													
	Grey-crowned Babbler *Pomatostomus temporalis*		C	●		●					●	●	●	●	●	C	●	●	●				●	●	●	●	●	●
	White-browed Babbler *Pomatostomus superciliosus*															UC	●						●	●	●		●	●
	Chiming Wedgebill	✧														UC	●			●	●						●	●

Chestnut Quail-thrush *Cinclosoma castanotus*															R	●			●	●						●	
Cinnamon Quail-thrush *Cinclosoma cinnamomeum*															UC	●			●	●						●	●
Varied Sitella *Daphoenositta chrysoptera*		UC	●		●				●	●	●	●	●	●	UC	●	●					●	●	●		●	●
Crested Shrike-tit *Falcunculus frontatus*		R	●										●	●													
Crested Bellbird *Oreoica gutturalis*															C	●		●	●			●	●	●	●	●	●
Mangrove Golden Whistler *Pachycephala melanura*	✧	UC		●		●			●		●			●													
Grey Whistler *Pachycephala simplex*		C		●		●			●	●	●	●	●														
Rufous Whistler *Pachycephala rufiventris*		C	●	●	●	●			●	●	●	●	●	●	C	●	●	●				●	●	●	●	●	●
White-breasted Whistler *Pachycephala lanioides*	✧	UC				●			●		●			●													
Little Shrike-thrush *Colluricincla megarhyncha*		UC		●		●		●	●	●	●	●	●	●													
Sandstone Shrike-thrush *Colluricincla woodwardi*	✧	UC			●						●	●	●	●													
Grey Shrike-thrush *Colluricincla harmonica*		UC	●		●				●	●	●	●	●	●	C	●	●	●				●	●	●	●	●	●
Broad-billed Flycatcher *Myiagra ruficollis*	✧	UC	●	●	●	●		●	●	●	●	●	●	●													
Leaden Flycatcher *Myiagra rubecula*		UC	●	●	●				●	●	●	●	●	●													
Shining Flycatcher *Myiagra alecto*		C		●	●	●		●	●	●	●	●	●	●													
Restless Flycatcher *Myiagra inquieta*		UC	●		●			●	●	●	●	●	●	●													
Magpie-lark *Grallina cyanoleuca*		C	●	●	●			●	●	●	●	●	●	●	C	●	●	●	●	●	●	●	●	●	●	●	●
Rufous Fantail *Rhipidura rufifrons*		UC		●		●			●	●	●	●	●	●													

Checklist			Top End Habitats							Top End Areas						Centre Habitats							Centre Areas					
Ticks	Common names and scientific names	Special Birds	Status	Woodland	Monsoon Forest	Escarpment	Mangroves	Mudflats & Beaches	Wetlands	Darwin	Small Parks	Kakadu	Litchfield	Katherine	VRD	Status	Shrubland	Riverine	Ranges	Sand Dunes	Plains	Wetlands	Alice Springs	East MacDonnells	West MacDonnells	Watarrka	Uluru	South East Desert
	Grey Fantail *Rhipidura fuliginosa*		R	●	●						●	●		●	●	R	●	●	●				●	●	●	●	●	●
	Mangrove Grey Fantail *Rhipidura phasiana*	✧	UC				●			●		●																
	Northern Fantail *Rhipidura rufiventris*		C	●	●	●	●		●	●	●	●	●	●	●													
	Willie Wagtail *Rhipidura leucophrys*		C	●		●			●	●	●	●	●	●	●	C	●	●	●	●		●	●	●	●	●	●	●
	Spangled Drongo *Dicrurus bracteatus*		CSM	●	●	●	●		●	●	●	●	●	●	●													
	Black-faced Cuckoo-shrike *Coracina novaehollandiae*		C	●	●	●	●		●	●	●	●	●	●	●	C	●	●	●	●		●	●	●	●	●	●	●
	White-bellied Cuckoo-shrike *Coracina papuensis*		C	●	●	●	●		●	●	●	●	●	●	●													
	Cicadabird *Coracina tenuirostris*		UC	●	●					●	●	●	●	●	●													
	Ground Cuckoo-shrike *Coracina maxima*	✧	UC												●	UCN	●	●			●		●	●	●	●	●	●
	White-winged Triller *Lalage sueurii*		C	●		●			●	●	●	●	●	●	●	CN	●	●	●				●	●	●	●	●	●
	Varied Triller *Lalage leucomela*		C	●	●	●	●		●	●	●	●	●	●	●													
	Yellow Oriole *Oriolus flavocinctus*		C	●	●	●	●		●	●	●	●	●	●	●													
	Olive-backed Oriole *Oriolus sagittatus*		UC	●	●	●	●		●	●	●	●	●	●	●													
	Figbird		C	●	●	●				●	●	●	●	●	●													

White-breasted Woodswallow *Artamus leucorynchus*	C	●		●	●	●	●	●	●	●	●	●	●	R	●	●				●	●	●	●		●	●
Masked Woodswallow *Artamus personatus*	RN	●						●	●	●	●	●	●	CN	●	●	●	●	●		●	●	●	●	●	●
White-browed Woodswallow *Artamus superciliosus*	RN	●						●	●	●		●	●	R	●							●	●		●	●
Black-faced Woodswallow *Artamus cinereus*	C	●						●	●	●	●	●	●	C	●	●	●	●	●		●	●	●	●	●	●
Little Woodswallow *Artamus minor*	C	●	●	●				●	●	●	●	●	●	UC			●				●	●	●	●	●	●
Black Butcherbird *Cracticus quoyi*	UC		●		●		●	●	●	●			●													
Grey Butcherbird *Cracticus torquatus*	UC	●		●				●	●	●	●	●	●	R	●						●	●	●	●	●	●
Pied Butcherbird *Cracticus nigrogularis*	C	●		●				●	●	●	●	●	●	C	●	●	●	●	●	●	●	●	●	●	●	●
Australian Magpie *Gymnorhina tibicen*	R	●										●	●	UC	●	●	●	●	●	●	●	●	●	●	●	●
Grey Currawong *Strepera versicolor*														Ex												
Australian Raven *Corvus coronoides*														R	●	●					●		●	●		●
Little Crow *Corvus bennetti*														UC	●	●	●	●	●	●	●	●	●	●	●	●
Torresian Crow *Corvus orru*	C	●	●	●	●	●	●	●	●	●	●	●	●	C	●	●	●	●	●	●	●	●	●	●	●	●
Apostlebird *Struthidea cinerea*	UC	●								●		●	●													
Western Bowerbird *Chlamydera guttata* ✧														C	●	●	●				●	●	●	●	●	●
Great Bowerbird *Chlamydera nuchalis*	C	●		●			●	●	●	●	●	●	●													
Singing Bushlark *Mirafra javanica*	UC	●					●	●	●	●	●	●	●	R					●						●	●
Richard's Pipit *Anthus novaeseelandiae*	C	●					●	●	●	●	●	●	●	C					●		●	●	●	●	●	●

Ticks	Common names and scientific names	Special Birds	Top End Habitats: Status	Woodland	Monsoon Forest	Escarpment	Mangroves	Mudflats & Beaches	Wetlands	Top End Areas: Darwin	Small Parks	Kakadu	Litchfield	Katherine	VRD	Centre Habitats: Status	Shrubland	Riverine	Ranges	Sand Dunes	Plains	Wetlands	Centre Areas: Alice Springs	East MacDonnells	West MacDonnells	Watarrka	Uluru	South East Desert
	Yellow Wagtail *Motacilla flava*		RSM						●	●	●	●																
	Grey Wagtail *Motacilla cinerea*		V			●			●																			
	White Wagtail *Motacilla alba*		V						●																			
	House Sparrow *Passer domesticus*		R											●														
	Zebra Finch *Taeniopygia guttata*		R	●						●	●			●	●	C	●	●	●	●	●	●	●	●	●	●	●	●
	Double-barred Finch *Taeniopygia bichenovii*		C	●	●	●	●		●	●	●	●	●	●	●													
	Long-tailed Finch *Poephila acuticauda*		C	●		●			●	●	●	●	●	●	●													
	Masked Finch *Poephila personata*		C	●		●			●	●	●	●	●	●	●													
	Crimson Finch *Neochmia phaeton*		C	●	●	●	●		●	●	●	●	●	●	●													
	Star Finch *Neochmia ruficauda*	✧	R	●					●	●	●	●		●	●													
	Painted Finch *Emblema pictum*	✧	R			●									●	UC	●	●	●					●	●	●	●	
	Yellow-rumped Mannikin *Lonchura flaviprymna*	✧	UC	●					●	●	●	●		●	●													
	Chestnut-breasted Mannikin *Lonchura castaneothorax*		UC	●					●	●	●	●	●	●	●													
	Pictorella Mannikin *Heteromunia pectoralis*	✧	R	●					●		●	●	●	●	●													

Gouldian Finch *Erythrura gouldiae*	✧	R	●								●	●	●	●												
Mistletoebird *Dicaeum hirundinaceum*		C	●	●	●	●		●	●	●	●	●	●	●	C	●	●	●				●	●	●	●	●	●
White-backed Swallow *Cheramoeca leucosternus*		V	●		●										UC	●	●	●	●	●	●	●	●	●	●	●	●
Barn Swallow *Hirundo rustica*		V					●			●	●																
Welcome Swallow *Hirundo neoxena*		V	●	●	●	●	●	●							UC						●					●	●
Tree Martin *Hirundo nigricans*		C	●	●	●	●	●	●	●	●	●	●	●	●	UC		●				●	●	●	●	●	●	●
Fairy Martin *Hirundo ariel*		UC	●	●	●	●	●	●	●	●	●	●	●	●	UC		●				●	●	●	●	●	●	●
Clamorous Reed-Warbler *Acrocephalus stentoreus*		R						●	●	●	●	●	●	●	UC						●	●		●			
Oriental Reed-Warbler *Acrocephalus orientalis*		V						●																			
Tawny Grassbird *Megalurus timoriensis*		UC						●	●	●	●	●	●	●													
Little Grassbird *Megalurus gramineus*															UC						●	●		●		●	
Spinifexbird *Eremiornis carteri*	✧														UC			●		●			●	●			
Rufous Songlark *Cincloramphus mathewsi*		UC	●					●	●	●	●	●	●	●	UC	●	●					●	●	●	●	●	●
Brown Songlark *Cincloramphus cruralis*		UC						●	●	●	●				UC	●	●			●		●	●	●	●	●	●
Zitting Cisticola *Cisticola juncidis*	✧	UC						●	●	●	●	●															
Golden-headed Cisticola *Cisticola exilis*		C	●					●	●	●	●	●	●	●	V						●	●					
Yellow White-eye *Zosterops luteus*		C		●		●		●	●	●	●			●													
Common Starling *Sturnus vulgaris*		V	●																								

APPENDIX 1 - USEFUL ADDRESSES

Birding Organisations
TREC Gondwana
Birdwatching Tours
PO Box 37
Alice Springs NT 0871
Phone (08) 8952 8248
Fax (08) 8952 8247
Email: nature@trec.com.au

Birds Australia (Royal Australasian Ornithologists Union)
415 Riversdale Road
Hawthorn East VIC 3123
Phone (03) 9882 2622

Bird Observers Club of Australia
183 Springvale Road
Nunawading VIC 3131
PO Box 185
Nunawading VIC 3131
Phone (03) 9877 5342

Environmental Organisations
Environment Centre of the Northern Territory
24 Cavenagh Street
Darwin NT 0800
PO Box 2120
Darwin NT 0801
Phone (08) 8981 1984

Arid Lands Environment Centre
Shop 3, Gregory Terrace
Alice Springs NT 0870
PO Box 2791
Alice Springs NT 0871
Phone (08) 8952 2497

Northern Territory Field Naturalists
PO Box 39565
Winnellie NT 0820
Phone (08) 8946 6467 or
(08) 8945 0805

Land Councils
(for the Centre)
Central Land Council
33 Stuart Highway
Alice Springs NT 0870
PO Box 3321
Alice Springs NT 0871
Phone (08) 8951 6320

(for the Top End)
Northern Land Council
9 Rowling Street
Casuarina NT 0810
PO Box 42921
Casuarina NT 0811
Phone (08) 8920 5178

Leave a couple of weeks for processing of permit applications.

Maps
Lands, Planning and Environment
Nichols Place
Corner Bennett and Cavenagh Streets
Darwin NT 0800
Phone (08) 8999 7032

Lands, Planning and Environment
Lands and Housing Building
21 Gregory Terrace
Alice Springs NT 0870
Phone (08) 8951 5756

Park Management Authorities
Parks Australia North
MLC Building, 81 Smith Street
Darwin NT 0800
GPO Box 1260
Darwin NT 0801
Phone (08) 8946 4300

Parks and Wildlife Commission of the Northern Territory

Darwin
Goyder Building
Palmerston NT 0830
PO Box 496
Palmerston NT 0831
Phone (08) 8999 4555

Alice Springs
Stuart Highway
Alice Springs NT 0870
PO Box 1046
Alice Springs NT 0871
Phone (08) 8951 8211

Katherine
Giles Street
Katherine NT 0850
PO Box 344
Katherine NT 0851
Phone (08) 8973 8888

Pastoral Leaseholders

Under Graziers in the Yellow Pages of the Northern Territory phone book or contact Lands, Planning and Environment, Darwin (08) 8999 4455, Alice Springs (08) 8951 5316

Road Conditions

(08) 8922 3394 (NT Police road report — recorded message)
Police Stations throughout the Territory

Tourism Organisations

Central Australian Tourism Industry Association
Centre Point Building,
Cnr Hartley Street and Gregory Terrace
Alice Springs NT 0870
PO Box 2227
Alice Springs NT 0871
Phone (08) 8952 5199

Darwin Region Tourism Association
33 Smith Street Mall
Darwin NT 0800
GPO Box 4392
Darwin NT 0801
Phone (08) 8981 4300

Katherine Regional Tourist Association
Cnr Stuart Highway and Lindsay Street
Katherine NT 0850
PO Box 555
Katherine NT 0851
Phone (08) 8972 2650

Tennant Creek Regional Tourist Association
Shop 1, Transit Centre
Patterson Street
Tennant Creek NT 0860
PO Box 601
Tennant Creek NT 0861
Phone (08) 8962 3388

NT Holiday Information Helpline
Phone 1800 62 1336

APPENDIX 2 - PLANT NAMES

Common Name	Scientific Name
Trees and Shrubs	
acacia species	*acacia* spp.
Bloodwood	*Eucalyptus opaca (formerly E. terminalis)*
bluebush species	*Maireana* spp. (formerly *Kochia* spp.)
cassia species	*Senna* spp. (formerly *Cassia* spp.)
chenopod species	saltbush, bluebush and samphire species & others
Coolibah	*Eucalyptus microtheca*
corkwood species	*Hakea eyreana & H.suberea*
cycad species	*Cycas* spp.
Cypress Pine (Top End)	*Callitris intratropica*
Desert Oak	*Allocasuarina decaisneana*
eremophila species	*Eremophila* spp.
eucalypt species	*Eucalyptus* spp.
Fern-leaved Grevillea	*Grevillea pteridifolia*
fig species	*Ficus* spp.
Freshwater Mangrove	*Barringtonia acutangula*
Ghost Gum	*Eucalyptus papuana*
gidgee species	*Acacia cambagei* and *A. georginae* & others
grevillea species	*Grevillea* spp.
gum trees	see eucalypt species
Hill Mulga	*Acacia macdonnelliensis*
hopbush species	*Dodonaea* spp,
Ironwood	*Acacia estrophiolata*
Lignum	*Muehlenbeckia cunninghamii*
mallee species	*Eucalyptus gamophylla, E. pachyphylla & E. socialis* & others
MacDonnell Ranges Cycad	*Macrozamia macdonnelliensis*
mangrove species	*Avicennia* sp., *Rhizophora* spp. & others
Mulga	*Acacia aneura*
Native Fig	*Ficus platypoda*
pandanus species	*Pandanus* spp.
paperbark species	*Melaleuca* spp.
Red Mulga	*Acacia cyperophylla*
River Red Gum	*Eucalyptus camaldulensis*
Salmon Gum	*Eucalyptus tintinnans*
saltbush species	*Atriplex* spp., *Enchylaena* sp. & *Rhagodia* spp. & others
samphire species	*Halosarcia* spp.
Stringybark	*Eucalyptus tetrodonta*
Waddy Wood	*Acacia peuce*
wattles	see acacia species
White Cypress Pine	*Callitris glaucophylla*
Whitewood	*Atalaya hemiglauca*
Witchetty Bush	*Acacia kempeana*
Woollybutt	*Eucalyptus miniata*
Grasses	
Mitchell grass species	*Astrebla* spp.
Sandhill Cane-grass	*Zygochloa paradoxa*
sorghum species	*Sorghum* spp.
spinifex species	*Triodia* spp. & *Plectrachne* spp.

APPENDIX 3 - FURTHER READING

Field Guides

Pizzey, G. *A Field Guide to the Birds of Australia.* Angus and Robertson, Sydney 1991.

Simpson and Day. *Field Guide to the Birds of Australia.* Penguin Books Australia Ltd, Ringwood 1996.

Slater, P., P. Slater & R. Slater. *The Slater Field Guide to Australian Birds.* Lansdowne Publishing Pty Ltd, Sydney 1989.

Other Bird Books

Blakers, M., S.J.J.F. Davies & P.N. Reilly. *The Atlas of Australian Birds.* Melbourne University Press, Carlton 1984.

Christidis, L., & W.E. Boles. *The Taxonomy and Species of Birds of Australia and Its Territories.* RAOU, Melbourne 1994.

Forshaw, J.M. & W.T. Cooper. *Australian Parrots.* Lansdowne-Rigby Publishers, Sydney 1981.

Frith, H.J. *Pigeons and Doves of Australia.* Rigby Publishers, Sydney 1982.

Garnett, S. *The Action Plan for Australian Birds.* ANPWS, Canberra 1992.

Hayman, P., J. Marchant, & T. Prater. *Shorebirds; An Identification Guide to the Waders of the World.* Christopher Helm Ltd, London 1986.

Hollands, D. *Eagles Hawks and Falcons of Australia.* Thomas Nelson Australia, Melbourne 1984.

Hollands, D. *Birds of the Night: Owls, Frogmouths and Nightjars of Australia.* Reed Books Pty. Ltd., Sydney 1991.

Marchant, S., & P.J. Higgins (Eds). *Handbook of Australian, New Zealand and Antarctic Birds. Volume 1: Ratites to Ducks.* Oxford University Press, Melbourne 1990.

Marchant, S., & P.J. Higgins (Eds). *Handbook of Australian, New Zealand and Antarctic Birds. Volume 2: Raptors to Lapwings.* Oxford University Press, Melbourne 1993.

Higgins, P.J., & S.J.J.F. Davies (Eds). *Handbook of Australian, New Zealand and Antarctic Birds. Volume 3: Snipe to Pigeons.* Oxford University Press, Melbourne 1996.

Reader's Digest. *Reader's Digest Complete Book of Australian Birds.* Reader's Digest (Australia) Pty. Ltd., Sydney 1990.

Plant Books

Brock, J. *Native Plants of Northern Australia.* Reed, Chatswood, NSW 1993.

Jessop, J. (ed). *Flora of Central Australia.* Reed Books Pty Ltd, Sydney 1981.

Urban, A. *Wildflowers & Plants of Central Australia.* Southbank Editions. Port Melbourne 1990.

APPENDIX 4 - ALTERNATIVE BIRD NAMES

Alexandra's Parrot (see Princess Parrot)
Australian Dotterel (see Inland Dotterel)
Australian Goshawk (see Brown Goshawk)
Australian Kestrel (see Nankeen Kestrel)
Australian Reed-Warbler (see Clamorous Reed-Warbler)
Banded Plover (see Banded Lapwing)
Beach Thick-knee (see Beach Stone-curlew)
Black Duck (see Pacific Black Duck)
Black Cormorant (see Great Cormorant)
Blue-winged Shoveler (see Australasian Shoveler)
Boobook (see Southern Boobook)
Broad-billed Roller (see Dollarbird)
Broad-tailed Thornbill (see Inland Thornbill)
Brown Whistler (see Grey Whistler)
Burdekin Duck (see Radjah Shelduck)
Bush Thick-knee (see Bush Stone-curlew)
Calamanthus (see Rufous Fieldwren)
Crested Hawk (see Pacific Baza)
Eastern Reef Heron (see Eastern Reef Egret)
Emblema (see Painted Finch)
European Starling (see Common Starling)
Feral Pigeon (see Rock Dove)
Flock Pigeon (see Flock Bronzewing)
Fork-tailed Kite (see Black Kite)
Gibber Chat (see Gibberbird)
Golden-backed Honeyeater (see Black-chinned Honeyeater)
Grass Whistling-Duck (see Plumed Whistling-Duck)
Green-backed Warbler (see Green-backed Gerygone)
Green-wing Pigeon (see Emerald Dove)
Happy Jack (see Apostlebird)
Indian Koel (see Common Koel)
Jabiru (see Black-necked Stork)
Land Mullet (see Blind Mullet)
Large Sand Plover (see Greater Sand Plover)
Large-billed Warbler (see Large-billed Gerygone)
Lavender-flanked Wren (see Variegated Fairy-wren)
Least Frigatebird (see Lesser Frigatebird)
Little Cuckoo-shrike (see White-bellied Cuckoo-shrike)
Little Falcon (see Australian Hobby)
Lotusbird (see Comb-crested Jacana)

Malay Bronze-Cuckoo (see Little Bronze-Cuckoo)
Mangrove Heron (see Striated Heron)
Mangrove Kingfisher (see Collared Kingfisher)
Mangrove Warbler (see Mangrove Gerygone)
Marsh Crake (see Baillon's Crake)
Marsh Harrier (see Swamp Harrier)
Masked Plover (see Masked Lapwing)
Mongolian Plover (see Lesser Sand Plover)
Mopoke (see Southern Boobook)
Mountain Duck (see Australian Shelduck)
Pacific Heron (see White-necked Heron)
Painted Firetail (see Painted Finch)
Paperbark Flycatcher (see Restless Flycatcher)
Pied Goose (see Magpie Goose)
Pied Stilt (see Black-winged Stilt)
Pink Cockatoo (see Major Mitchell's Cockatoo)
Plumed Egret (see Intermediate Egret)
Plumed Tree-duck (see Plumed Whistling-Duck)
Port Lincoln Parrot (see Australian Ringneck)
Red-collared Lorikeet (see Rainbow Lorikeet)
Reef Heron (see Eastern Reef Egret)
Reeve (see Ruff)
Rufous Night Heron (see Nankeen Night Heron)
Rufous Shrike-thrush (see Little Shrike-thrush)
Sandstone Friarbird (see Helmeted Friarbird)
Sacred Ibis (see Australian White Ibis)
Spotted Crake (see Australian Spotted Crake)
Spurwinged Plover (see Masked Lapwing)
Torresian Imperial Pigeon (see Pied Imperial-Pigeon)
Torres Strait Imperial-Pigeon (see Pied Imperial-Pigeon)
Variable Goshawk (see Grey Goshawk)
Wandering Tree-duck (see Wandering Whistling-Duck)
Warbler (see Gerygone)
Water Whistling-Duck (see Wandering Whistling-Duck)
Western Warbler (see Western Gerygone)
White-breasted Sea-Eagle (see White-bellied Sea-Eagle)
White-eyed Duck (see Hardhead)
White Goshawk (see Grey Goshawk)
White Ibis (see Australian White Ibis)
White-rumped Miner (see Yellow-throated Miner)
White-throated Warbler (see White-throated Gerygone)
Yellow-fronted Honeyeater (see Grey-fronted Honeyeater)

APPENDIX 5 - PLACE INDEX